THE STARGAZER'S COMPANION

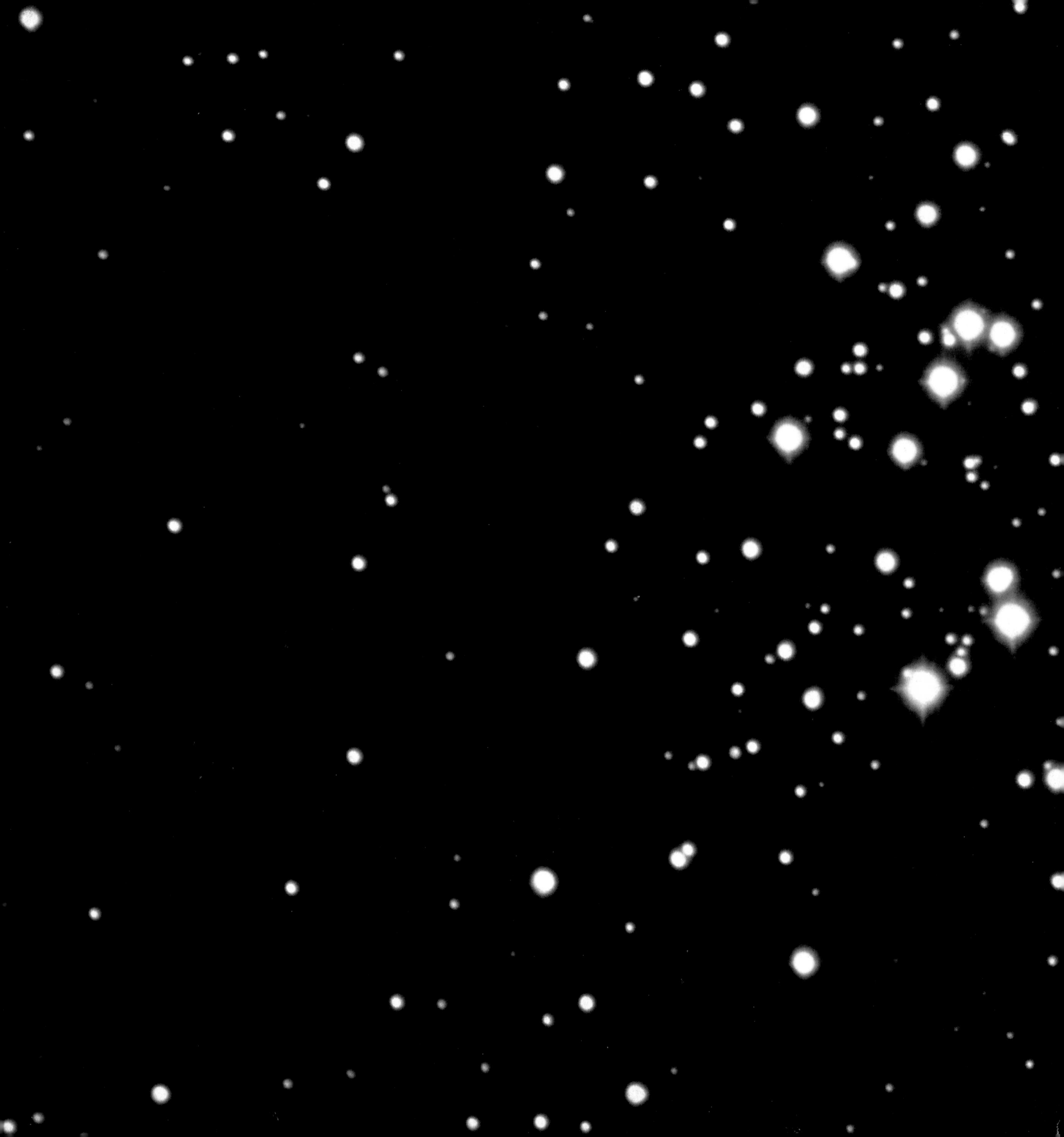

THE STARGAZER'S COMPANION

JAMES K. BLUM

MALLARD PRESS
An imprint of BDD Promotional Book Company, Inc.
666 Fifth Avenue
New York, New York 10103

A FRIEDMAN GROUP BOOK

Published by MALLARD PRESS
An imprint of BDD Promotional Book Company, Inc.
666 Fifth Avenue
New York, New York 10103

First published in the United States of America in 1990 by The Mallard Press.

ISBN 0-792-45262-3

THE STARGAZER'S COMPANION
was prepared and produced by
Michael Friedman Publishing Group, Inc.
15 West 26th Street
New York, New York 10010

Editor: Sharyn Rosart
Art Director: Jeff Batzli
Designer: Edward Noriega
Photo Researcher: Daniella Jo Nilva
Production: Karen L. Greenberg
Production Editor: Suzanne DeRouen

Typeset by: Interface Group, Inc.
Color separations by Kwong Ming Graphicprint Co.
Printed and bound in Hong Kong by South China Printing Co.

TABLE OF CONTENTS

This galaxy contains billions of stars, many thousands of which are just like our own sun and may contain systems of planets similar to our own.

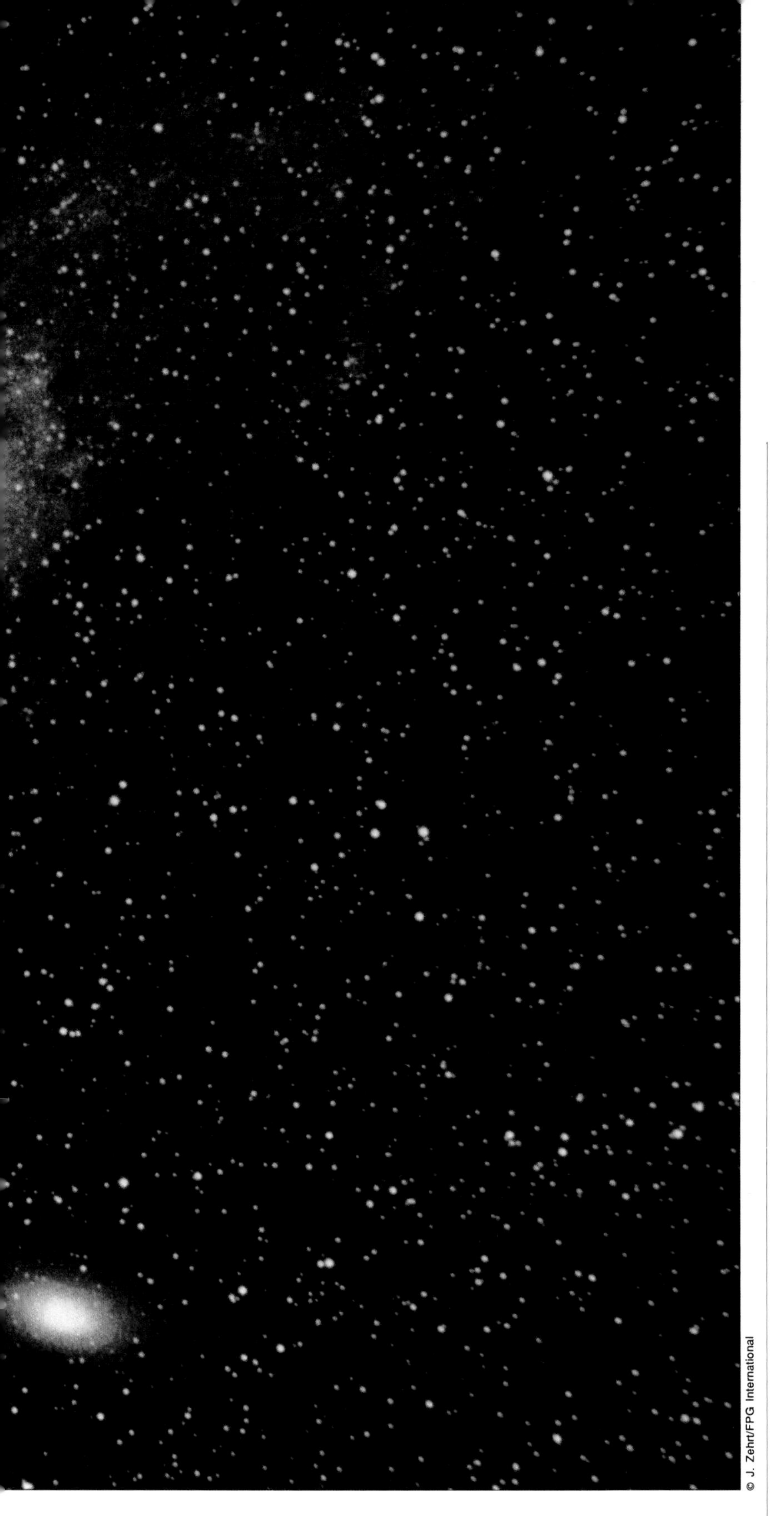

© J. Zehrt/FPG International

INTRODUCTION

The night sky is an arena full of activity, open to all free of charge. On any night with reasonable visibility, a novice stargazer —using no equipment other than a portable reclining lawn chair and a very basic star map—may view one or more comets, a meteor shower, the auroras, several planets, and hundreds or even thousands of stars. Add a pair of low-power field glasses or binoculars and, depending upon the aperture of the lenses, the number of visible objects in the night sky may increase to as many as one hundred thousand. Exchange the field glasses for an eight-inch telescope and suddenly the sky becomes so densely packed with activity that it takes a very knowledgeable and diligent stargazer to navigate the celestial sea.

Being able to see several million stars certainly sounds enticing; however, for all but the truly dedicated and studious observers of the universe, finding a particular star or object among so many may be more trouble than viewing it is worth. Indeed, some backyard astronomers become discouraged and unhappy when they find that they aren't able to see some of the things they thought their telescope would show them. Too much aperture may be hazardous to a stargazer's enjoyment of the night sky, and enjoyment is really the key to stargazing. There is much for the stargazer to learn, and even a few things to be memorized; but if little or no fun is involved in the learning process, astronomy can become tedious awfully fast.

This is why *The Stargazer's Companion* was written. It is intended to be an aid to amateur astronomers, both beginners and more experienced observers. It is for anyone who has an interest in the beauty and wonder of the celestial sphere. If it serves its purpose, it will spark the curiosity of newcomers to stargazing; for those with an advanced interest, *The Stargazer's*

The density of stars in the night sky is sometimes overwhelming. Here, a nebula is clearly visible even through the veil of stars.

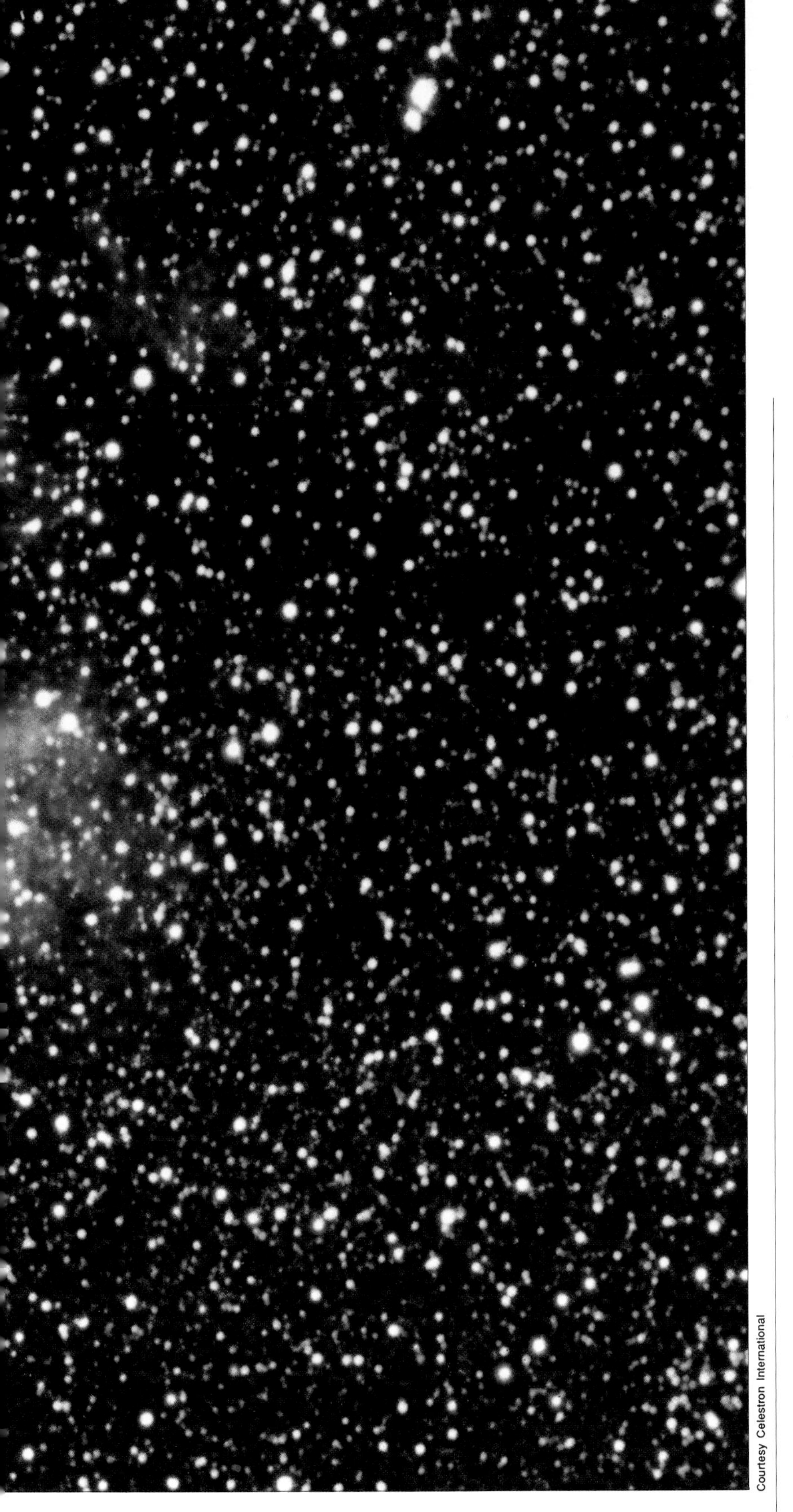

Courtesy Celestron International

Companion can illuminate new facets of one of the world's most popular leisure activities and perhaps aid in developing and refining an amateur astronomer's knowledge and ability.

There are two reasons why people look skyward: to expand their knowledge of the universe and the physical world and to enjoy the pleasure of looking at the stars. Though some astronomers feel the only reason to look through a telescope is to be "in service to Science" (as one popular astronomy book trumpets), *The Stargazer's Companion* is geared to the simple pleasure of watching the stars. An amateur sky observer may decide that he or she wishes to contribute regular observations to one of the many specialized astronomical observing bodies, such as the Association of Amateur Variable Star Observers (AAVSO) or the Association of Lunar and Planetary Observers (ALPO); information about how to get involved in these and other amateur observing organizations can be found in these pages. However, the best reason to look at the stars, and the reason that will keep most stargazers looking for years to come, is the absolute wonder and beauty of the night sky.

Various sections of this book offer information on how to look at the stars, what to look for, what equipment to use, and some of the history of stargazing. *The Stargazer's Companion* also contains descriptions of some of the more unusual and interesting galactic and interstellar phenomena, updated information on new advances in astronomy, and spectacular astrophotography.

Most importantly, *The Stargazer's Companion* is a book to help even beginners enjoy the pleasures of stargazing. Whether a stargazer's tastes run to discovering comets or tracking space junk, *The Stargazer's Companion* will kindle and refine the reader's interest in astronomy. Here's to good seeing!

C H A P T E R   O N E

A BRIEF HISTORY OF STARGAZING

© George East

This stunning nebula, seen also on page 124, is formed of gas and dust, probably from an exploded star. Nebulas, though often confused with galaxies by novice observers, are in fact the remnants of a star in a galaxy.

Although it is not necessary to know the history of astronomy to appreciate the pleasures of stargazing, some knowledge of stargazing's past can only enhance an amateur astronomer's sense of enjoyment and accomplishment. Knowing that a standard pair of 8 × 50 binoculars offers more optical precision and light-gathering ability than the instruments Galileo used to make his groundbreaking observations can add a sense of satisfaction to any stargazer's activity. However, the roots of astronomy are not based in science. In fact, astronomy's precursor is the pseudoscience of astrology, which grew out of human attempts to explain natural phenomena in mystical terms.

THE EARLIEST STARGAZERS

It's clear that even before humanity began to keep records, people watched the skies. Indeed, it's difficult to imagine people *not* looking skyward. What the first humans saw in the sky could not have been much different from what modern stargazers see. In fact, the constellations probably haven't changed significantly for as long as humans have inhabited the earth. Though we have no record of what earliest civilizations may have viewed in the night sky, we can imagine their amazement at the sight of the stars moving above them.

The early Chaldeans, who lived in the triangular region of Mesopotamia where the Tigris and Euphrates rivers meet, were the first to write down their observations of the heavens. Though they had few measuring instruments, they kept detailed records of the motions of the sun, the moon, the stars, and the five planets visible with the naked eye. They divided the stars into at least fifty-two constellations, including the twelve zodiacal constellations still used today in celestial reckoning.

Despite their ability to predict solar and lunar eclipses with some accuracy, the Chaldeans reacted to these occurrences with fear. They did not draw any scientific construction from their observations, but believed the stars and the planets, and particularly the sun and moon, to be deities able to control events on Earth. From these religious beginnings grew the deification of the celestial bodies throughout the pre-Christian era and the rise of the zodiacal pantheon.

The ancient Egyptians shared the Chaldeans' belief in the sky

ABOVE, a Mayan stone carving of a two-headed dragon symbolizes the supernatural violence that the Mayans felt accompanied the period when Venus appeared near to the sun in the sky. AT LEFT, the dragon symbol is again used, in an engraving from the mid-fifteenth century depicting the sun and moon being dragged through the sky by a huge dragon. Nearly every pre-Renaissance culture believed that the sky was the domain of deities, often depicted as huge and powerful beasts.

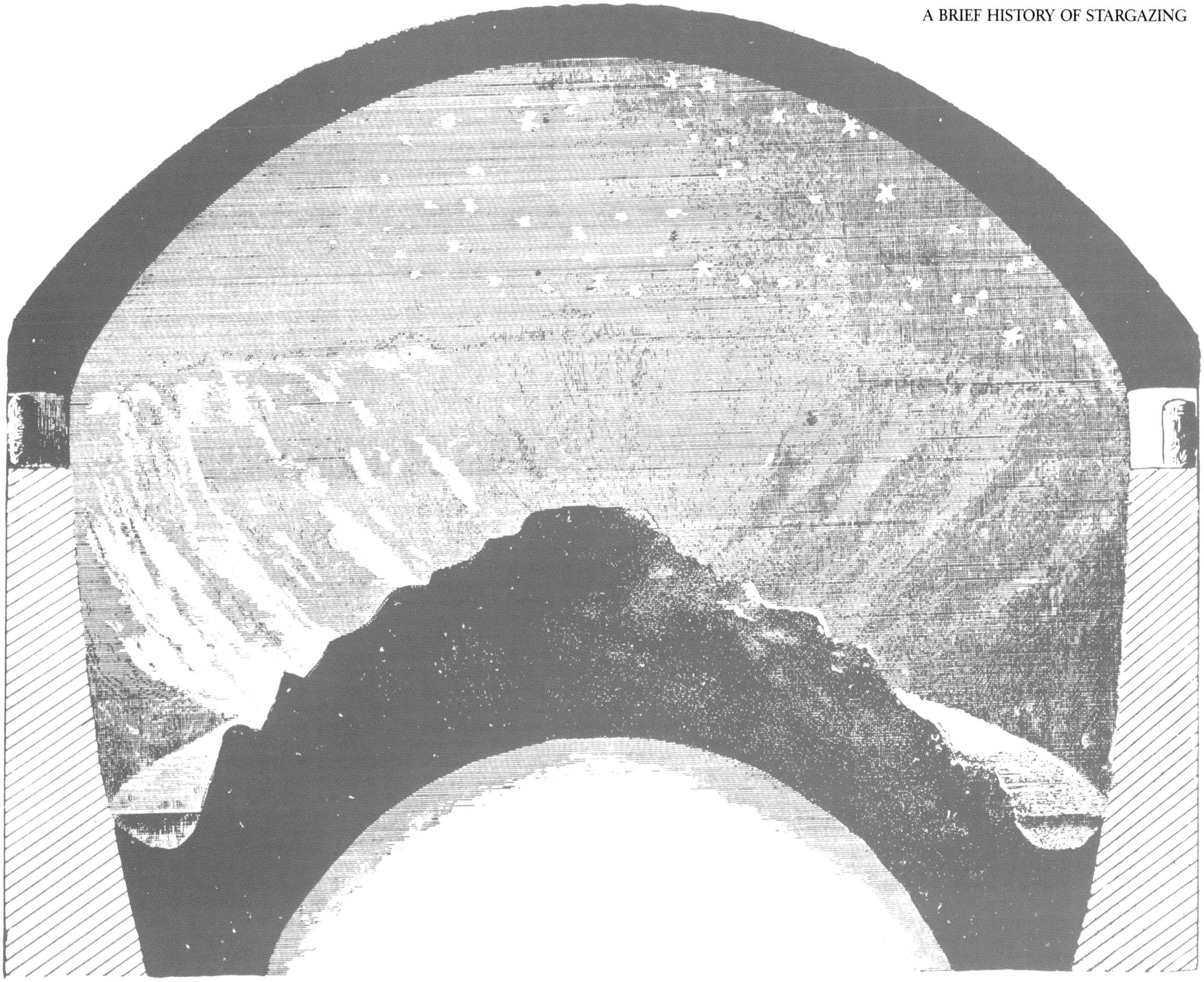

North Wind Picture Archive

The Chaldean view of the universe, as pictured above, maintained that the stars were affixed to a dome that was suspended over the Earth.

gods. However, they were less interested in measuring the movement of celestial bodies. In fact, though much has been written about the wonders of the ancient Egyptian civilization, its only real accomplishment in the field of stargazing was the development of a reasonably accurate calendar, a feat also accomplished by the Chaldeans and the ancient Chinese.

The Chinese's knowledge of the heavens apparently was at least as extensive as that of the Chaldeans. Unfortunately, modern historians are unable to accurately assess ancient Chinese astronomical knowledge: One Chinese emperor, furious with the inability of his astrologers to predict the future, had all of China's astronomy books burned in 213 B.C.

Despite their accurate observations, none of these ancient civilizations attempted to explain any of the events they witnessed in the night sky as other than the workings of the gods. None developed astronomy as a science; instead they used their observations to foretell the future and explain the behavior of the celestial deities they had created.

Courtesy NASA

THE BIRTH OF ASTRONOMY

Though extensive information about the cosmos had been collected by earlier civilizations, it was not until the height of classical Greece that this information began to be treated in a rational, scientific manner. The Greeks were the first to attempt to explain why natural events occurred without attaching a mystical component. It eventually became clear to them that their astrological beliefs did not correspond with the "rules" of the universe that were beginning to show themselves to Greek thinkers. By the sixth century B.C., belief in the Panhellenic gods had largely died out. Instead, these ancient Greeks made a concerted effort to comprehend and explain the physical universe in rational terms.

For nine hundred years, Greek philosophers were the world's scientists. The philosophers, so called because they favored logical reasoning over empirical problem solving, changed the course of humanity with their methods. Thales, first of the great Greek philosophers, believed that through observation and consideration of the heavens, all things of the universe could be known. He traveled extensively in the known world and brought the knowledge and records of the Chaldeans and Egyptians back to Greece. A century later, Pythagoras postulated that all things could be understood as functions of their numerical properties—sizes, distances, and ratios. Thus began the qualification of observations, the heart of astronomy today.

Ironically, it was the Greek fascination with numbers and geometric forms that also hampered the growth of astronomy for nearly two thousand years. Plato, believing that the movements of celestial bodies could be explained as circular orbits, formulated an extensive cosmology, or explanation of the workings of the universe. In Plato's incarnation, Earth was the center of the universe. The sun, the five planets visible with the naked eye, and the "fixed stars," which the Greeks believed to be points of light affixed to a transparent dome, moved around Earth in circular orbits. Plato also taught his students that the secrets of the universe could be deduced by thought alone. Because of Plato's teachings, the usual methodology in astronomy for nearly two millenia was to theorize about the universe and then make observations to support the theory, rather than the reverse.

This image of a new star, or protostar (see red patch at arrow) was created from infrared data produced by the Infrared Astronomical Satellite (IRAS). The star, in the constellation Perseus, is much like our sun was in its early stages, more than 4.5 billion years ago.

Aristotle, Plato's student, shared Plato's views and wrote extensively in support of Plato's cosmology. He also added many of his own theories and mathematical discoveries. His contribution to astronomy, like Plato's, was enormous; unfortunately, scholars and scientists—and even the Church—accorded these two thinkers such latitude that their errors were accepted as fact until the middle of the Renaissance. In fact, when Aristarchus of Samos proposed a theory that Earth and the visible planets revolved around the sun, it was scorned as being in direct conflict with Aristotle's laws of motion and dismissed for hundreds of years.

In the second century B.C., Hipparchus dedicated himself to making detailed observations to support Aristotle's theories. Among his many accomplishments are the measurement of the moon's period of orbit, exact tables of the motions of the moon and sun (it was still believed that the sun moved around the Earth), and, most importantly, a catalogue of over one thousand stars, categorized into six levels of brightness. This scale of star magnitudes formed the basis of the magnitude scale still in use today. Hipparchus's contributions were the last major astronomical achievements by the Greeks for three hundred years.

ABOVE, this cosmology represents a commonly held view of the universe before Galileo. In it, the sun and the five planets known to the ancients revolved around the Earth, with the twelve constellations of the zodiac anchoring the sphere of fixed stars. AT RIGHT, depending upon your observation point, you may be able to see the Great Nebula in Orion with no optical aid. The close proximity of the Great Nebula to Earth makes it a favorite object for stargazers. FOLLOWING PAGE, Nebulas such as this one, the North American Nebula, are made up of great clouds of gas and dust that have been dispersing over many thousands of years.

Ptolemy, in the second century A.D., was the last great Greek astronomer. His *Mathematical Collection* is an encyclopedia of all of the astronomical knowledge of his time, based on an Earth-centered cosmology and including his own extensive measurements and discoveries. This work, which comes to modern scholars by way of its Arabic translation, *Almagest* (Arabic for "The Greatest"), was the major astronomical textbook for scientists for fourteen centuries.

The Greek philosophers accomplished an amazing amount in less than a millenium, and their work would remain unequaled for many hundreds of years to come. They are generally held responsible for the birth of all the sciences, including astronomy. Though there were significant errors in several of their theories, the Greek philosophers advanced astronomy to such a point that it was not to be significantly improved until just prior to the development of the telescope.

Religion and Astronomy

Because religious belief in most cultures sprang up as a method of explaining natural phenomena, astronomy has always been intimately connected to religion. In many widely diverging ancient cultures, celestial objects came to be given distinct personalities and power over daily human life.

From the earliest civilizations came the original zodiacal astrology. The early Celts, the Mayans, the Persians, and scores of others had forms for celestial worship. Even as late as the seventeenth century, Christians believed that some celestial events, such as eclipses and supernovas, indicated that God was angry with Earth's inhabitants. It is no wonder, then, that today many people read horoscopes (which are published daily in thousands of newspapers in North America), believing that somehow the movements of the stars and planets have some effect on their everyday lives.

North Wind Picture Archive

ABOVE, Nicolaus Copernicus, the first to champion the theory of a sun-centered universe. Copernicus' observations, along with Tycho's extensive cataloguing of the stars, made possible the confirmation of this theory less than 100 years after it was initially put forward.

THE GOLDEN AGE OF ASTRONOMY

After Ptolemy, advances in astronomy ended for some time. The Romans, though they had access to all of the knowledge of the Greeks, made no significant discoveries. After the Romans, the Catholic church stifled astronomical research in the Western world for a thousand years. The Dark Ages produced no advances in any of the sciences. It was left to the Chinese and the Arabs, who inherited from the Greeks a fascination with the stars, to continue the study of astronomy. The Chinese are mainly responsible for the bulk of astronomical data available during the first thousand years after the birth of Christ. The Arabs took up the mantle from there, and their records cover the eleventh century and beyond.

However, real advances were not made in the Western world until the middle of the European Renaissance. Fifty years after Columbus discovered the Americas and made the case for a round Earth, Nicolaus Copernicus, a Polish mathematician and astronomer, showed that his observations were more comprehensible in a sun-centered universe than in an Earth-centered universe. Though this was a direct break with the teachings of the Catholic church at the time, it was not considered a serious offense until it began to gain widespread acceptance, many years after Copernicus' death in 1543.

Tycho Brahe, a Dane, took up the study of astronomy where Copernicus had left off, though he continued to believe in a geocentric universe. One of the most colorful figures in the history of astronomy, Tycho was also the first modern observational astronomer. Like Hipparchus before him, Tycho recorded thousands of meticulous observations and catalogued 777 stars. His assistant, Johannes Kepler, was the first to find conclusive evidence that the sun was at the center of the solar system (even in the early sixteenth century, it was believed that the universe consisted of only the objects in our solar system and a sphere of fixed stars). Kepler extrapolated from the observations made by his master and deduced that Earth revolved around the sun. He published several works on the geometry of celestial bodies, showed that the orbits of those bodies must be elliptical rather than circular, and formulated three basic laws of planetary motion. This work set the stage for the development of the telescope.

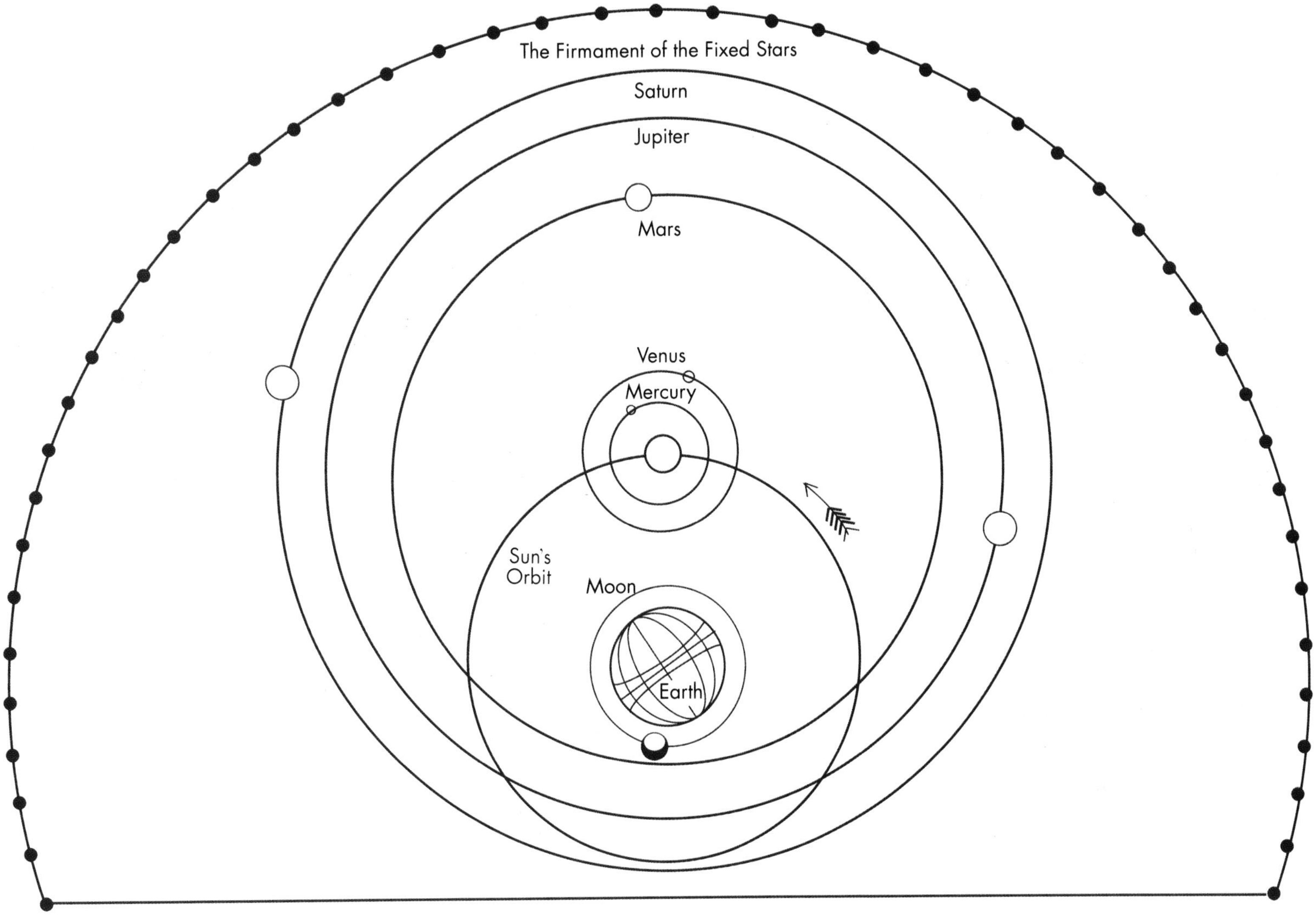

North Wind Picture Archive

In Tycho's cosmology, Mercury and Venus revolve around the sun, which in turn revolves around the Earth, while at the same time the moon and the other planets revolve around the Earth under the sphere of unmoving stars. The sheer impossibility that all of these bodies could move within several paths of revolution and never run into each other makes this system fanciful at best.

Tycho Brahe

Tycho Brahe is one of the most colorful characters in the history of astronomy. Although his scientific accomplishments earned him fame, any story of his life would be incomplete unless it mentioned his fiery personality.

Tycho was involved in a duel early in his twenties in which his nose was severed. Thereafter, his face was dominated by a huge copper prosthetic device in the shape of a nose.

He was given a tract of land on the island of Hven on which to build an observatory. Tycho boasted that Uranieborg, the name he gave to his observatory, cost the king a hundred tons of gold. However, Tycho was a tyrannical landlord, and was eventually run off the land by his angry tenants.

An apocryphal story about Tycho concerns the manner of his death. After lunching at the royal table, he had to relieve himself during a lengthy speech by the king. He knew he couldn't leave the table during the king's speech, and became more and more agitated as the king rambled on, until he literally reached the breaking point. Tycho's bladder burst and he suffered internal infections that led to his death two months later.

The alluring moon, the one object that never fails to capture the attention of would-be stargazers, is perhaps best viewed not in a telescope, but through binoculars. INSET, a drawing, circa 1848, of a large reflecting telescope in Melbourne, Australia.

THE AGE OF THE TELESCOPE

Until the seventeenth century, astronomical mathematics had advanced much further than observational astronomy. Despite the improvements Tycho made to Ptolemy's measuring devices, the instruments of astronomy remained crude. However, in 1609, Galileo began assembling small telescopes in Italy, based upon the models of Hans Lippershey. It appears that Galileo was able to use Lippershey's designs to make telescopes that were comparable in focal length and aperture to some of today's binoculars. With his early, simple instruments, Galileo was able to make the firsthand observations that revolutionized astronomy. He viewed the phases of Venus, four of the moons of Jupiter, and craters on our own moon. He saw the first visible proof of the difference between stars and planets (through a telescope, planets resolve into objects, while stars remain points of light, proving that stars are considerably farther away from Earth than the planets).

At the same time, Galileo was experimenting with the laws of physics. He set out to research the motions of bodies, and began

North Wind Picture Archive

North Wind Picture Archive

North Wind Picture Archive

In the pantheon of great astronomers, the three above are perhaps the greatest. From left, Sir Isaac Newton, William Herschel, and Galileo.

by making observations without previously formulated conclusions in mind. From his collected data, he devised the laws of uniform motion and uniformly accelerated motion, as well as the principle of inertia. Galileo's approach was directly opposed to the Platonic-Aristotelian method of pure thought and immediately replaced the Greek method as the basis of scientific observation.

Near the end of the seventeenth century, Isaac Newton wrote *Principia Mathematica*, which introduced the law of universal gravity, the final modernizing principle of both astronomy and physics. Newton, like Galileo, was one of the few extraordinary scientists able to make both practical and intellectual contributions to science. Newton succeeded in separating light waves with a prism and produced a new type of telescope, which bears his name and remains the most popular today.

With the exception of Einstein's discovery of the principle of relativity, astronomical advances since Newton have been limited primarily to the refinement of our knowledge and the study of the distant past and potential futures of the universe.

William Herschel is known as the modern father of astronomy. His painstaking observations and discoveries of the planet Neptune and the existence of binary stars are his best known accomplishments. Herschel's example of studious observational technique continues to affect astronomy today.

Modern astronomy and astrophysics is primarily concerned with probing ever deeper into the universe. Astronomers are making observations with bigger and better equipment, and new fields of astronomy such as radio astronomy, X-ray astronomy, and other nonvisible observations of the universe have sprung up and added to our knowledge of the universe. Modern theorists such as Stephen Hawking (whose book, *A Brief History of Time*, made cutting-edge astrophysics accessible to the general public) are attempting to formulate a Grand Unified Theory, or GUT, of the universe. A GUT will, supposedly, explain the workings of the entire universe, known and unknown. Others are working on defining the boundaries of the universe, attempting to explain newly discovered objects such as quasars, or searching for black holes, which have only been theorized thus far.

Is a GUT possible? Perhaps. But some astronomers feel that there is no theory that can adequately explain everything. One thing is sure: Astronomers will never stop trying to see to the end of the universe and explain all that they see. Human destiny will move ever outward, away from Earth.

Edmond Halley, INSET, and two of his beloved comets. No matter which direction the comet is traveling, invariably the tail is pointing away from the sun. This is caused by the effect of solar wind on the comet.

Newton's Patron

Sir Isaac Newton is responsible for many of the great scientific discoveries of all time. Yet today, his name may be less well known than that of his patron, Edmond Halley. Halley's contributions to astronomy are twofold. First and foremost, he edited Newton's *Principia Mathematica* and had it published at his own expense. He was, by turns, Newton's stern critic and devout follower.

Halley also indelibly linked himself to astronomy by undertaking serious study of comets. His name is attached to one of the major periodic comets, and it was indeed Halley who realized that this comet was the same one that had been sighted at regular intervals many times in history. During Halley's day, comets were popularly regarded as evil omens flung randomly from heaven. Halley's extensive research on the orbits of periodic comets revised popular notions of comets and is the basis for all work on comets since.

Though he was a minor figure in astronomy compared with Newton, it is ironic that while Newton's name will be forever synonymous with an apple, Halley will always be associated with one of the most spectacular of celestial visitors, the comet.

Courtesy Celestron International

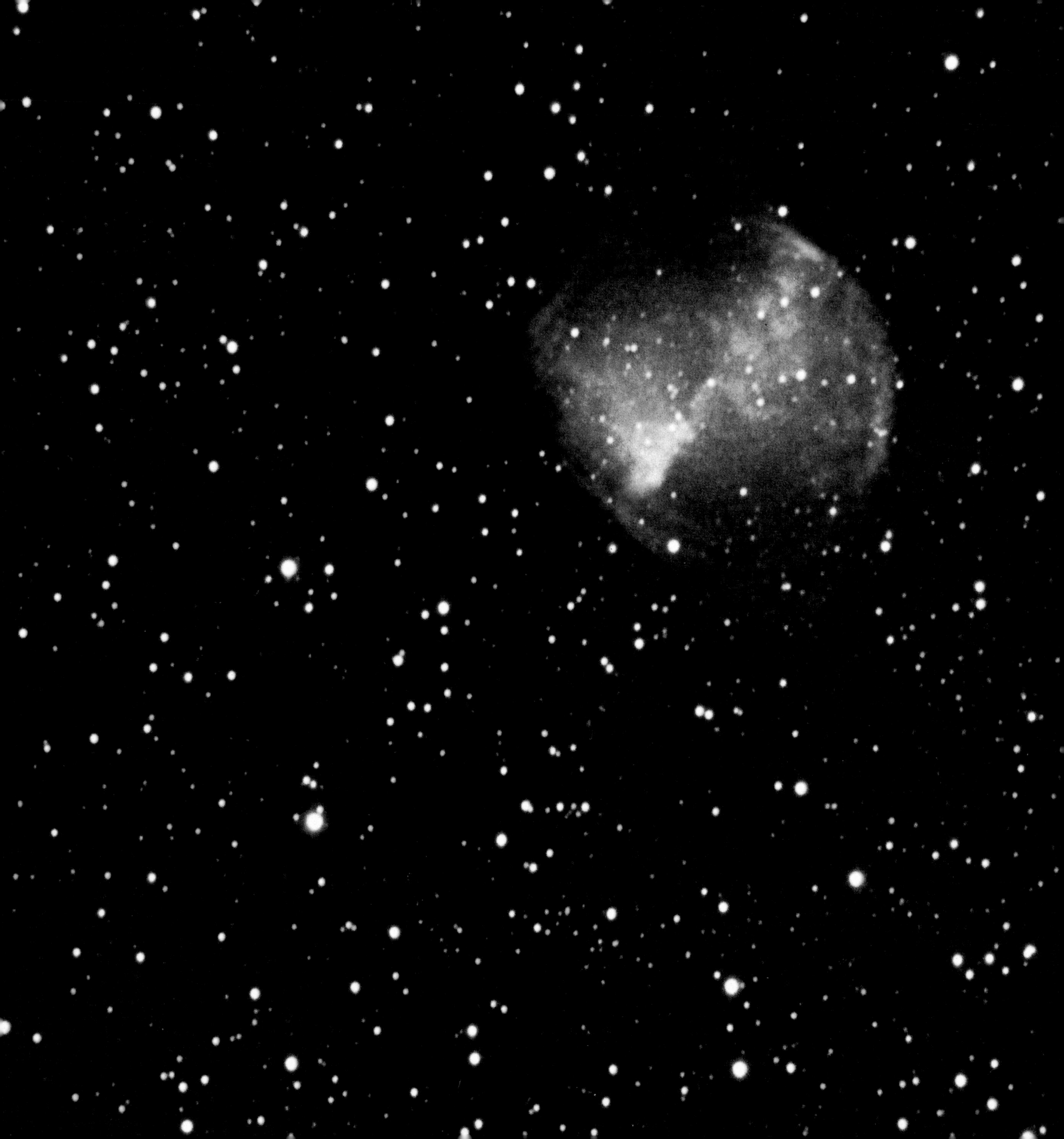

CHAPTER TWO

STARGAZING BASICS

Courtesy Celestron International

The Dumbbell Nebula is somewhat more distant than the Great Nebula in Orion, and requires a fairly large telescope to see.

THE CONSTELLATIONS

One of the best systems of memorization was developed thousands of years ago, originally by the Chaldeans in Mesopotamia: They divided the sky into constellations, or arbitrary patterns of stars. Until 1930, the constellations were very informal groupings of stars that, though widely used, were never specifically identified. That is, people knew and understood where, say, Sagittarius was in the sky, but there was no line of demarcation to say exactly where the field of Scorpius ended and that of Sagittarius began, and some of the lesser constellations were subject to debate as to their member stars. In 1930, the International Astronomical Union (IAU) specifically defined the constellations, giving each a precise boundary so that they could henceforth be used as the markers for distinct areas of the sky by astronomers. Although there are eighty-eight constellations recognized by the IAU, it isn't necessary for the novice stargazer to memorize all of them. The twelve zodiacal constellations form an excellent basis for beginning to learn the regions of the sky. Once these are familiar, other constellations and celestial objects can be learned easily.

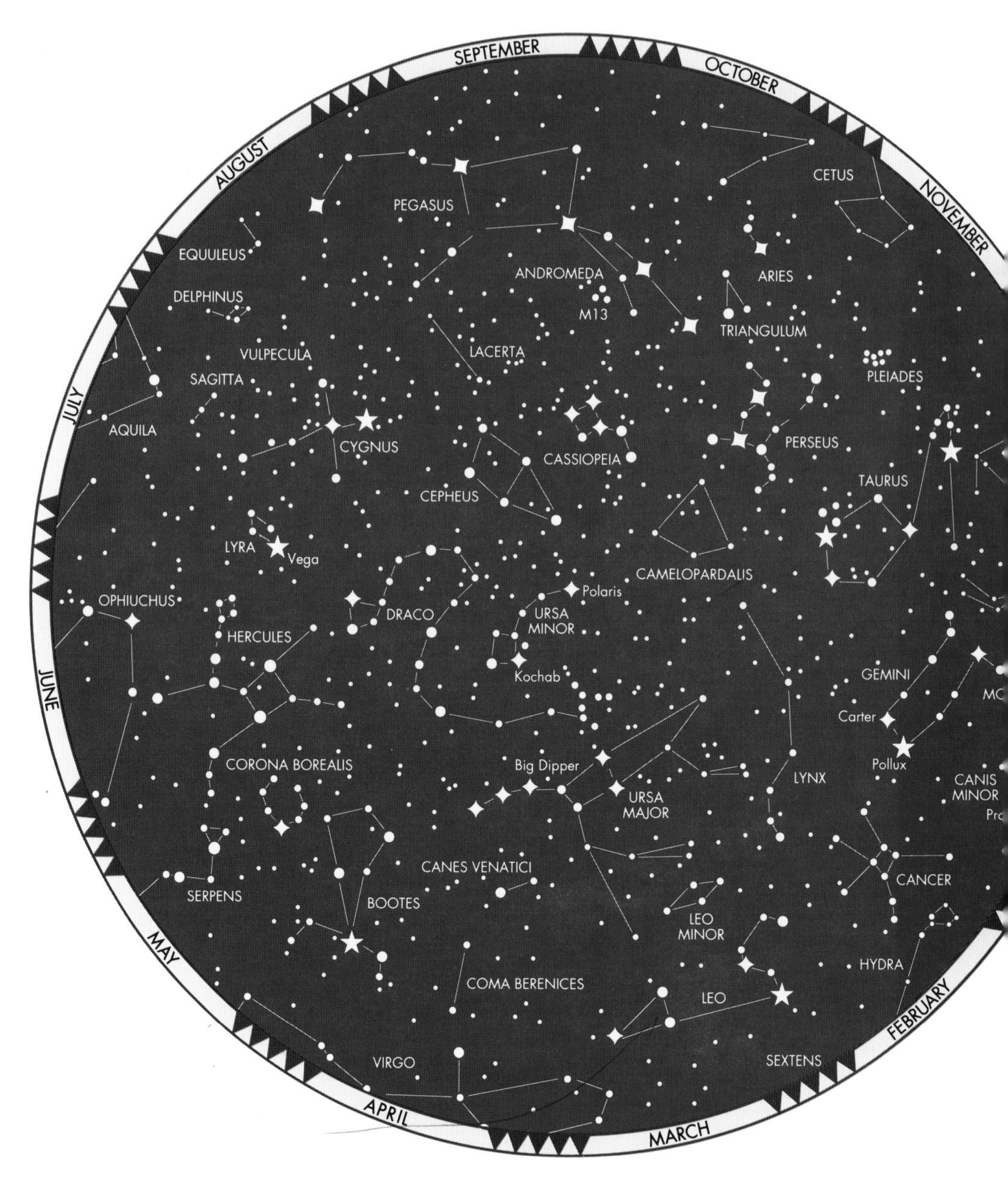

These diagrams show the corresponding positions of the stars of the Northern and Southern Hemispheres for each month, as seen from the North and South Poles, respectively. To correctly place the diagram, turn it until the current month is facing you. Diagrams such as these are available for all Earth latitudes.

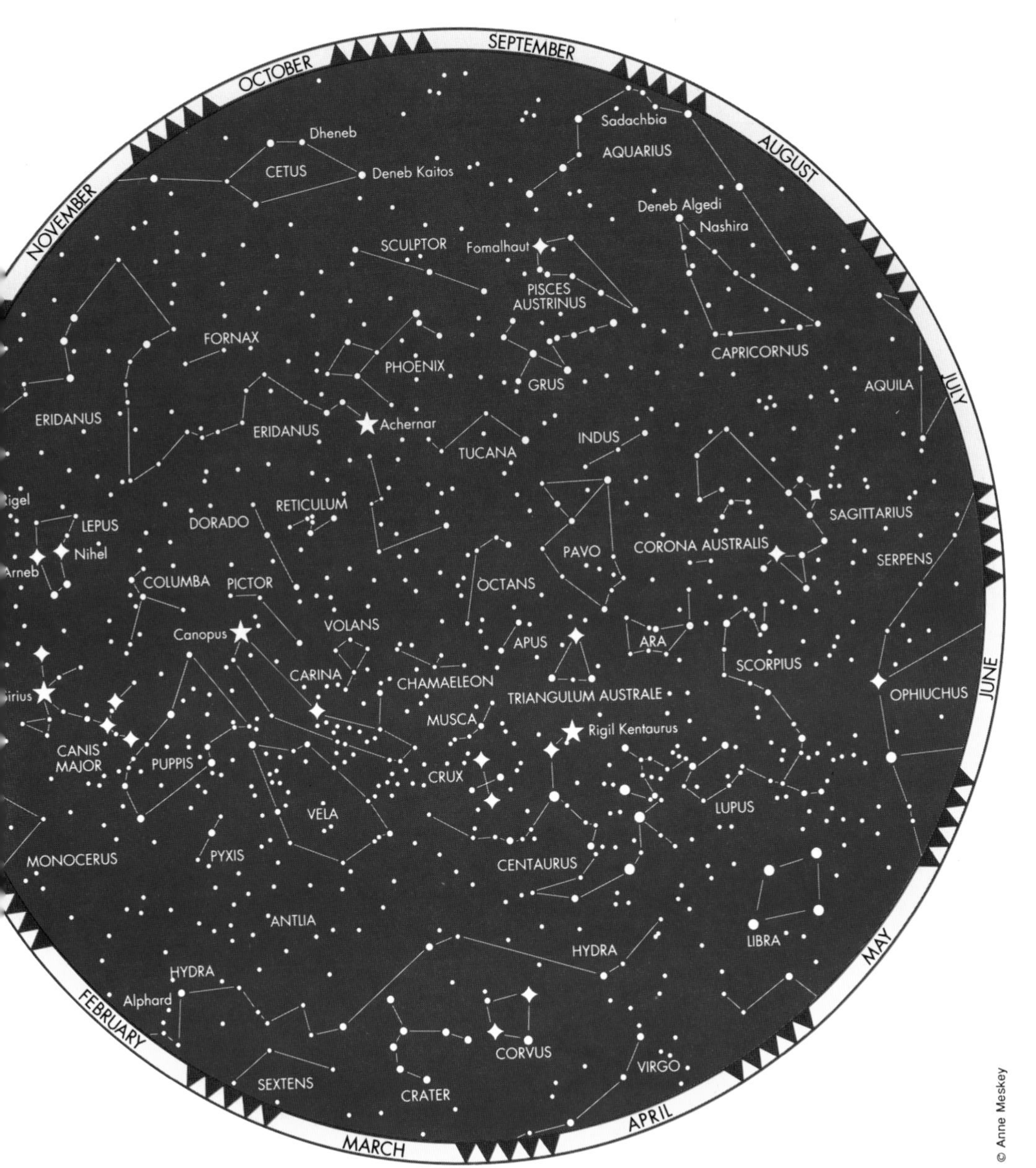

HOW TO FIND CELESTIAL BODIES

When you look at the sky, what do you see? Assuming it's a clear night, you are treated to the spectacular sight of a dark dome speckled with thousands of pinpoints of light. Yes, you know those points are stars (or perhaps planets or other celestial phenomena), but to make any sense of what you're seeing, you have to have some way of identifying the objects in the sky. Except for the Cardinal Rule of Skywatching—never look directly at the sun without eye protection—there is nothing more important for any stargazer, beginner or expert, to learn than the relative positions of the stars in the night sky.

With perhaps a thousand stars visible on any given evening, it sounds like a daunting task to learn the location of each of them. Of course, over thousands of years of stargazing, many systems and schemes for finding celestial bodies have been invented. However, only two have proven to be useful: separating sections of the sky into groups or patterns of stars called *constellations* and assigning to the sky a system of *celestial coordinates* (just as coordinates of latitude and longitude have been assigned to the surface of the earth). Each method has its strong and weak points, but mastering both methods is invaluable for any stargazer.

A star chart of the spring sky, with fanciful drawings of four of the best-known constellations, Gemini (the Twins), Virgo (the Virgin), Leo (the Lion), and Bootes (the Herdsman). As you can see from the diagram in the drawing of Leo, the positions of the stars in a constellation often don't coincide with an artist's conception of the constellation.

SPRING

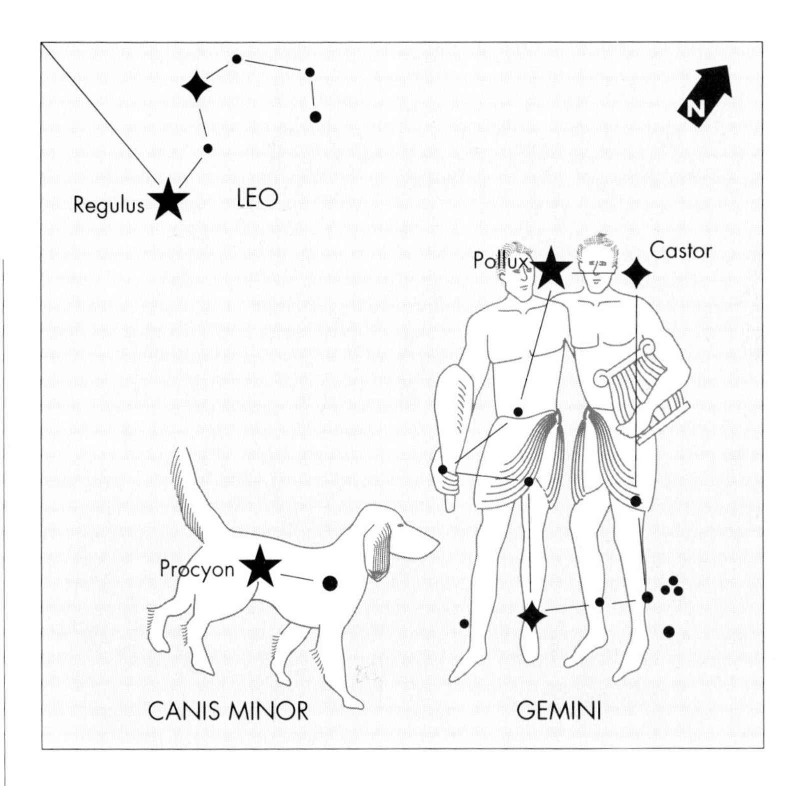
LEO
Regulus
N
Pollux
Castor
Procyon
CANIS MINOR
GEMINI

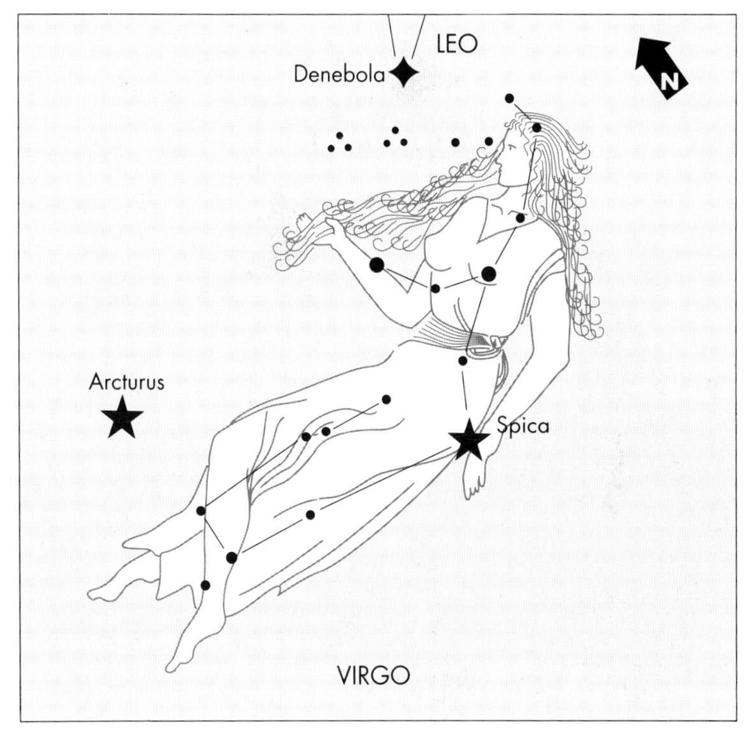
LEO
Denebola
N
Arcturus
Spica
VIRGO

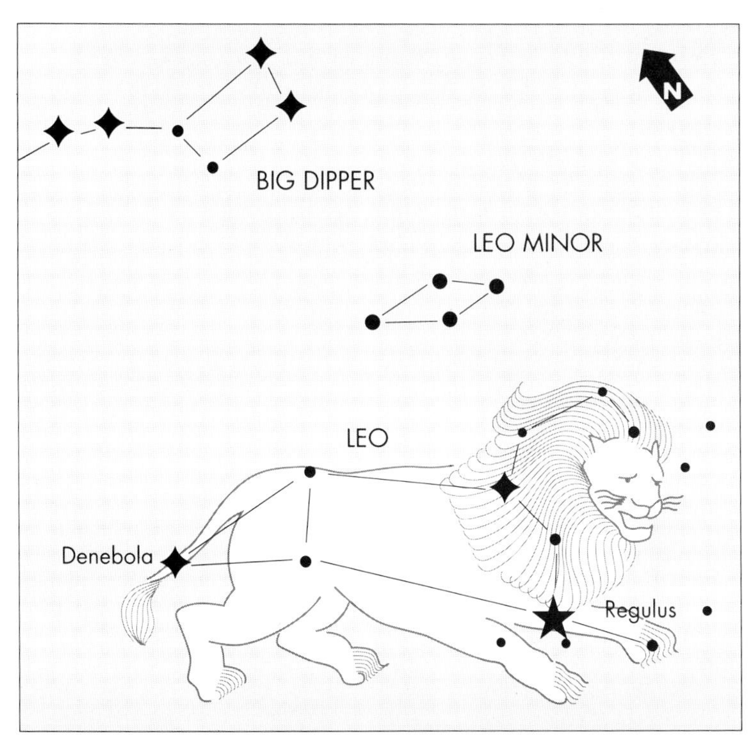
N
BIG DIPPER
LEO MINOR
LEO
Denebola
Regulus

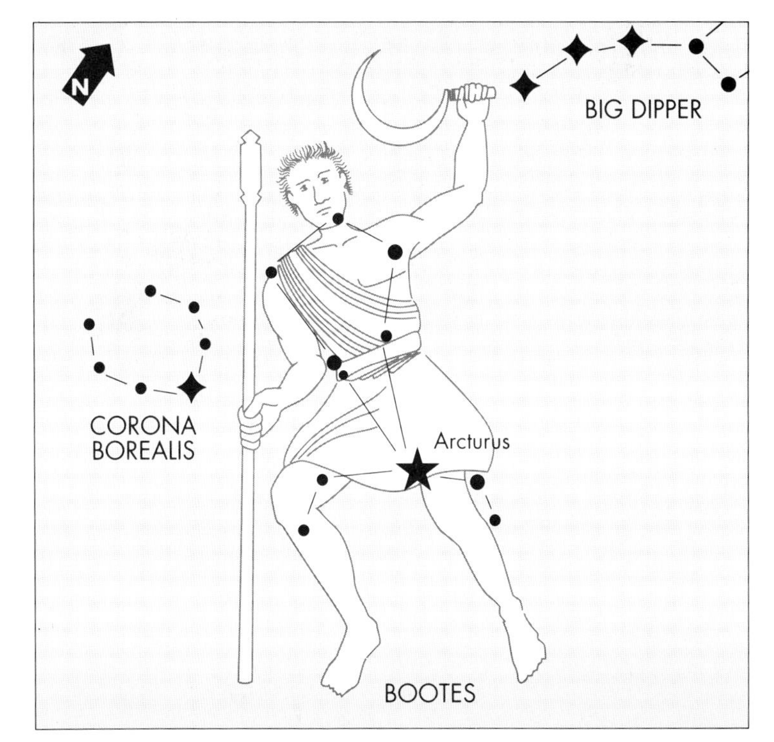
N
BIG DIPPER
CORONA
BOREALIS
Arcturus
BOOTES

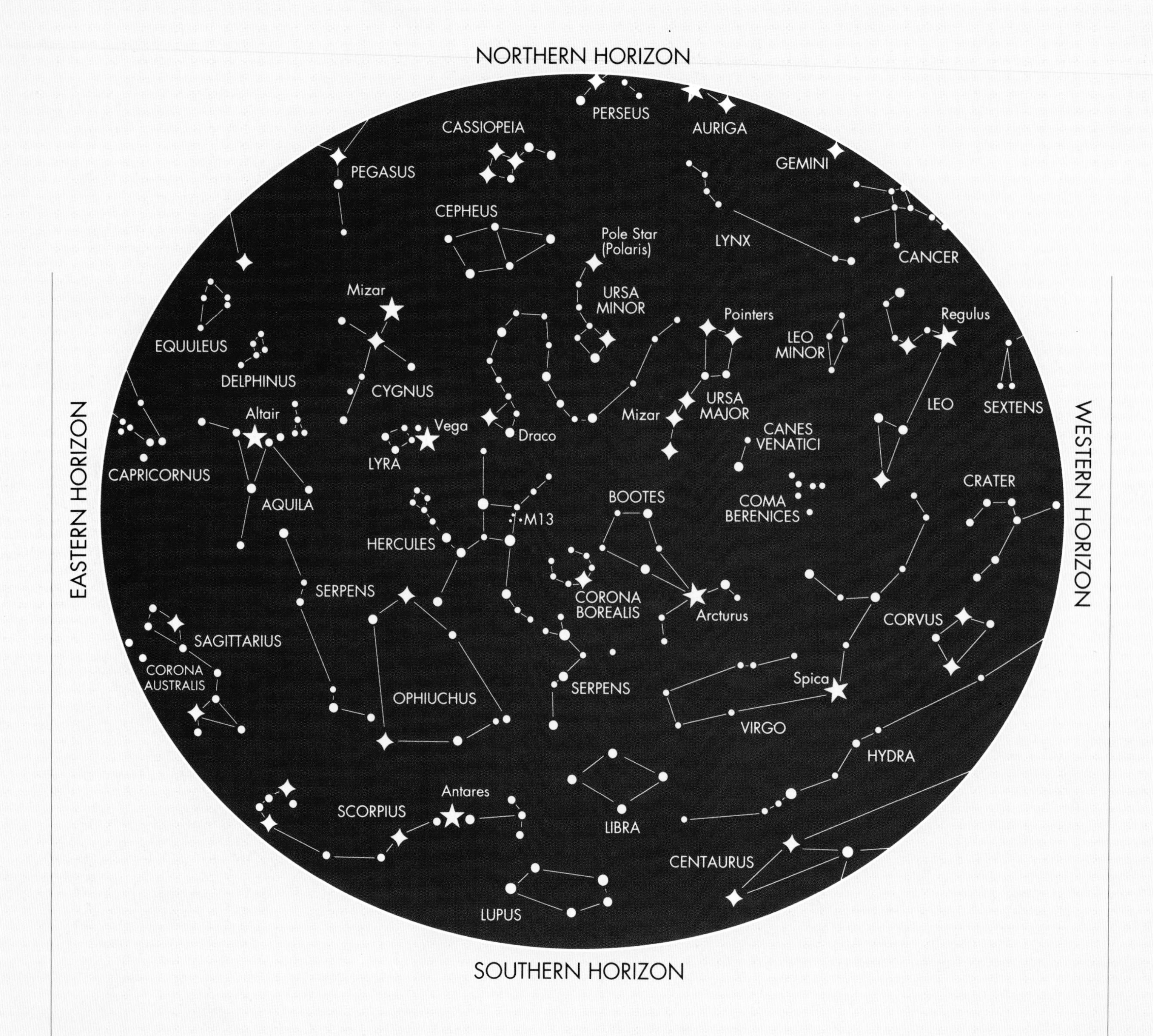

The stars of the summer sky are shown here, with representations of the constellations Lyra (the Lyre), Scorpius (the Scorpion), Hercules (the Strong Warrior), and Sagittarius (the Archer).

SUMMER

SUMMER

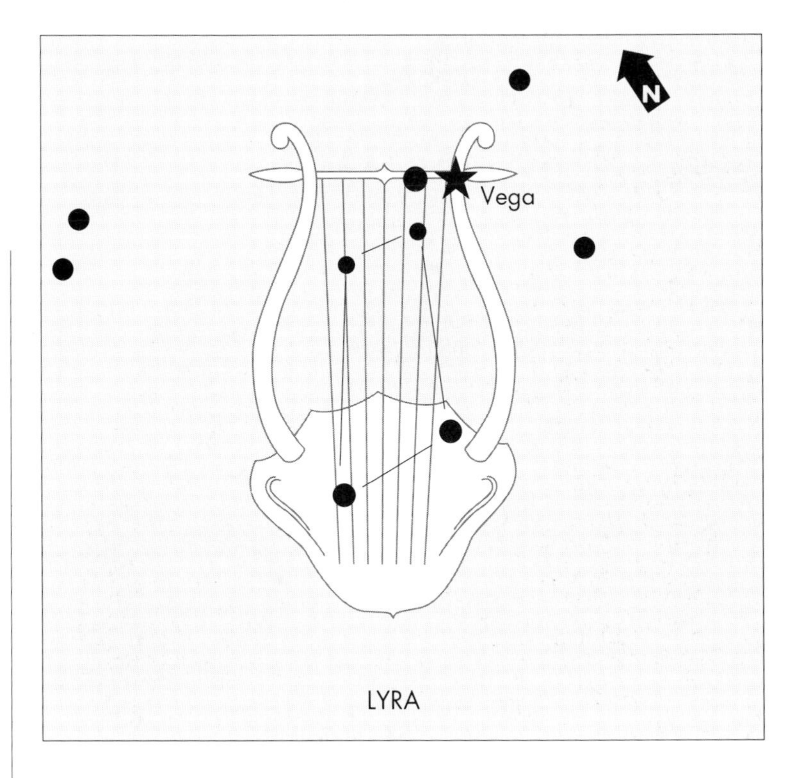
N
Vega
LYRA

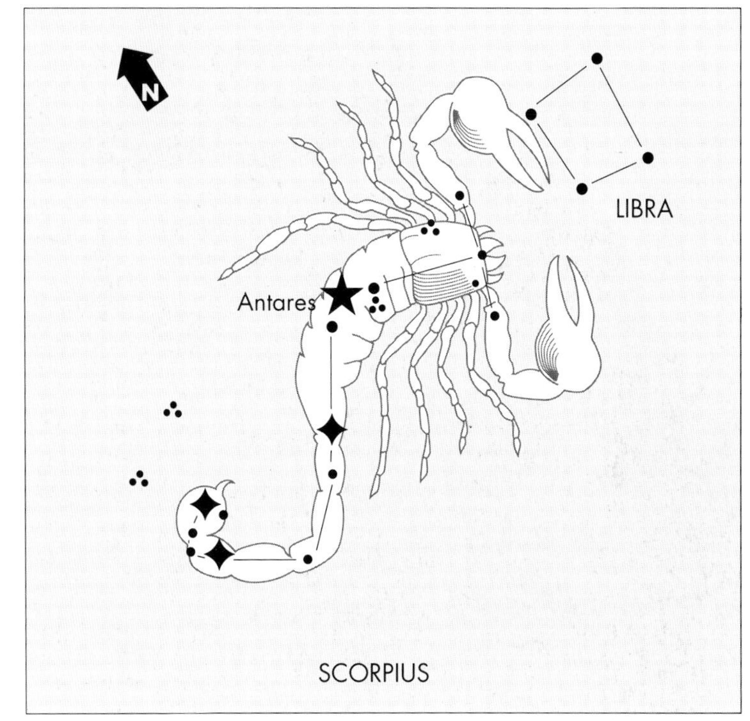
N
LIBRA
Antares
SCORPIUS

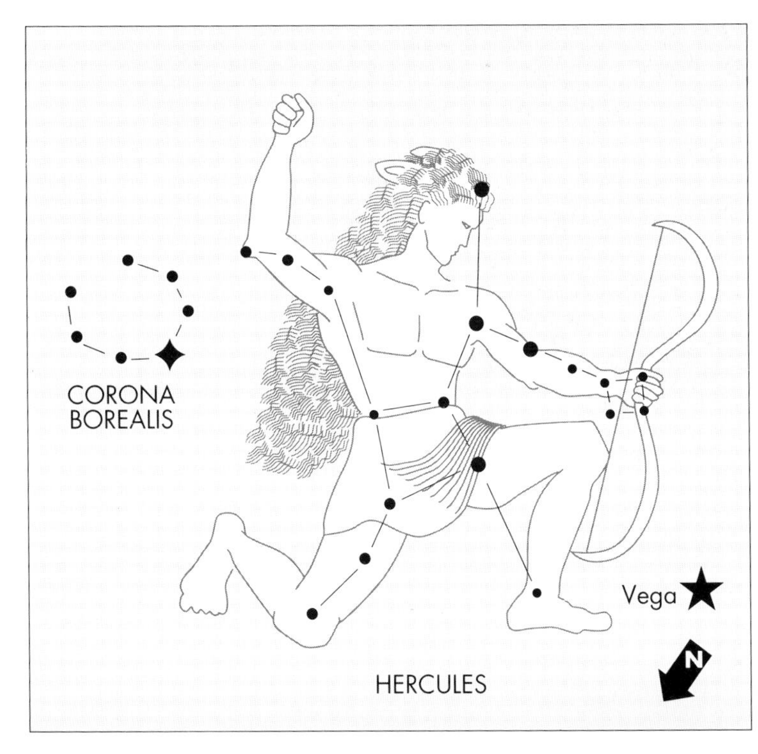
CORONA
BOREALIS
Vega
N
HERCULES

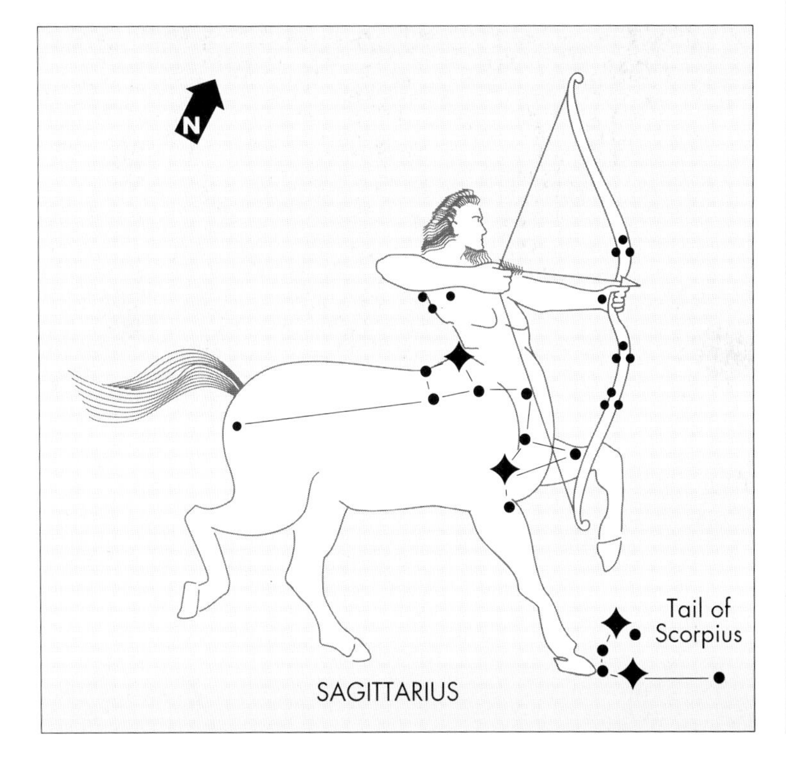
N
Tail of
Scorpius
SAGITTARIUS

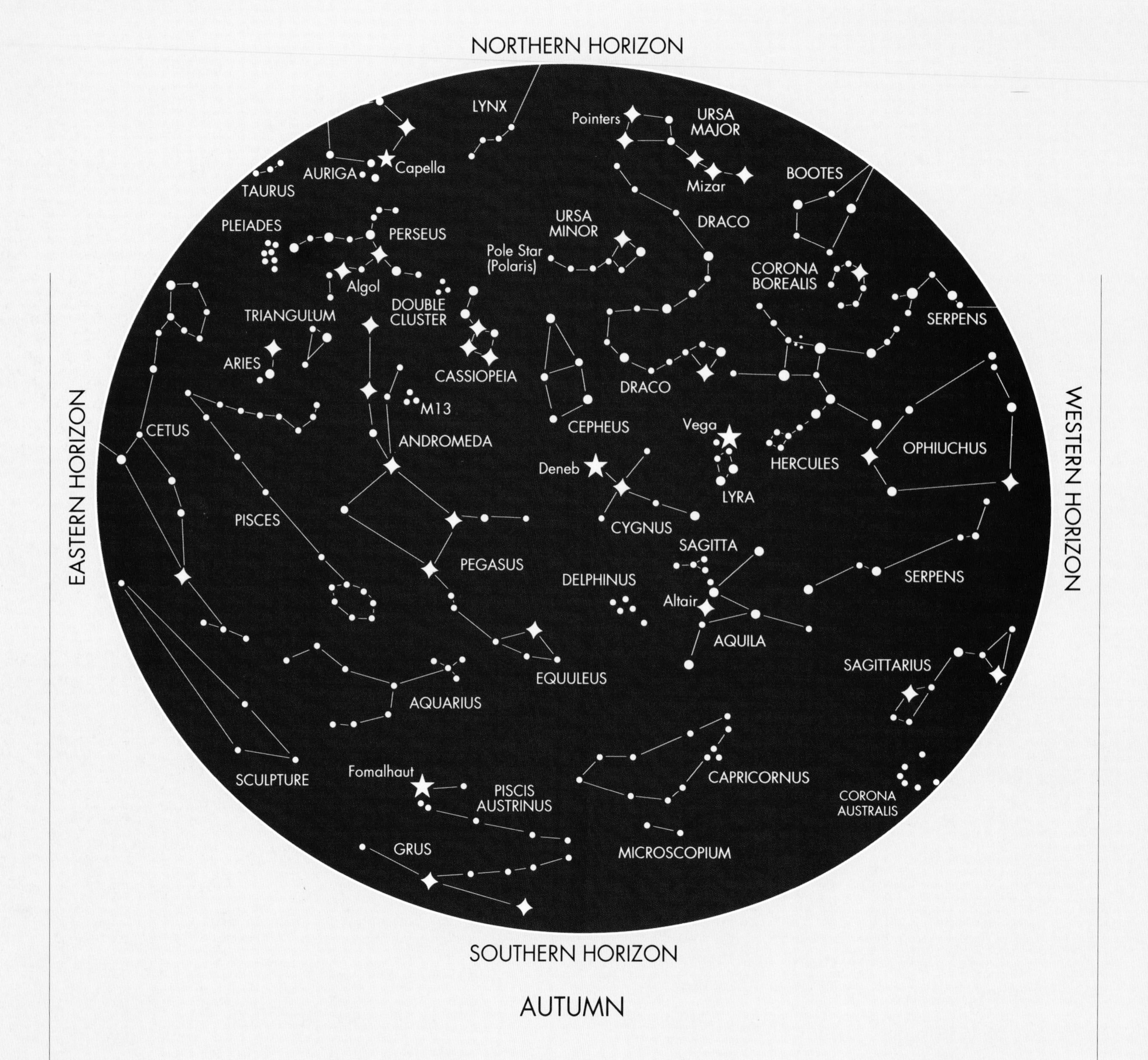

This star chart of the autumn sky is accompanied by illustrations, from left, of Perseus (the Hero), Andromeda (the Chained Princess), Auriga (the Charioteer), and Pegasus (the Winged Horse).

AUTUMN

CASSIOPEIA
PERSEUS
N

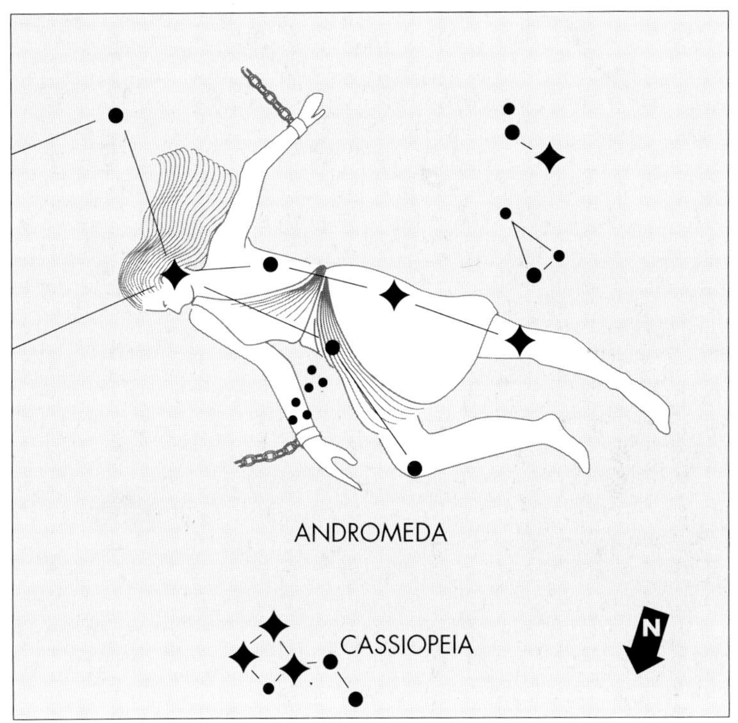
ANDROMEDA
CASSIOPEIA
N

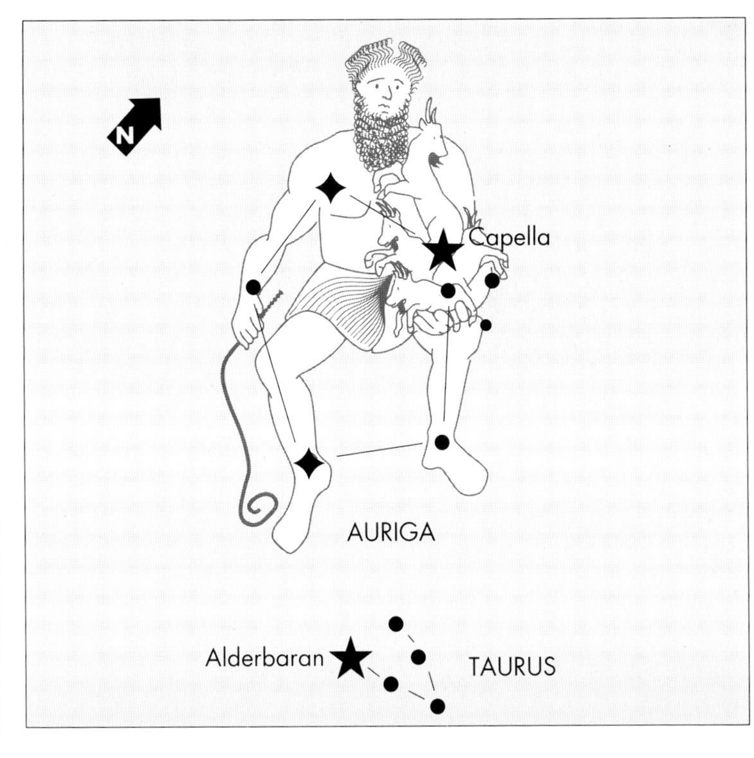
N
Capella
AURIGA
Alderbaran
TAURUS

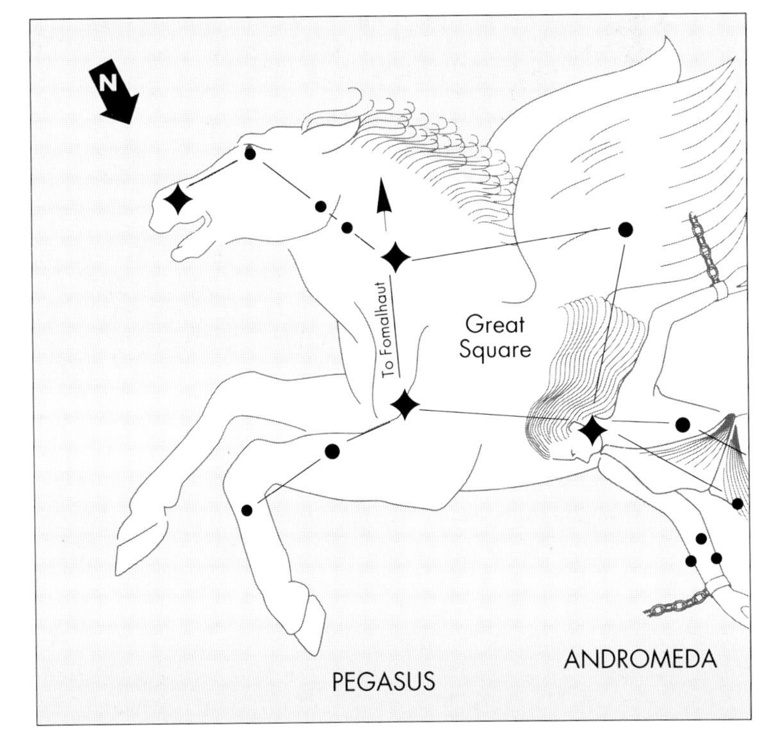
N
To Fomalhaut
Great
Square
PEGASUS
ANDROMEDA

WINTER

A star chart for the winter sky, with illustrations of Lepus (the Hare), Orion (the Hunter), Canis Major (the Great Dog), and Taurus (the Bull).

WINTER

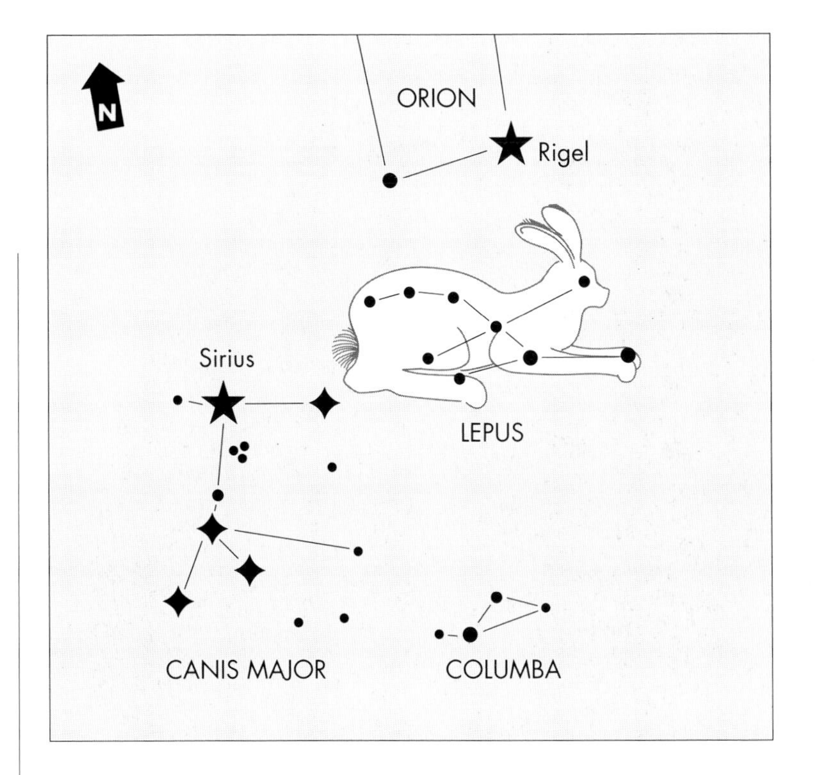
N
ORION
Rigel
Sirius
LEPUS
CANIS MAJOR
COLUMBA

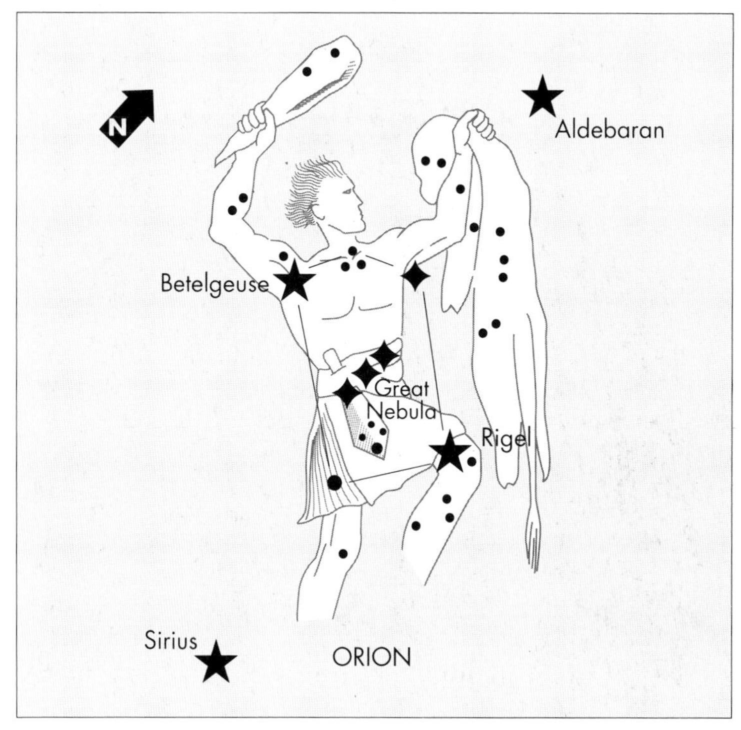
N
Aldebaran
Betelgeuse
Great
Nebula
Rigel
Sirius
ORION

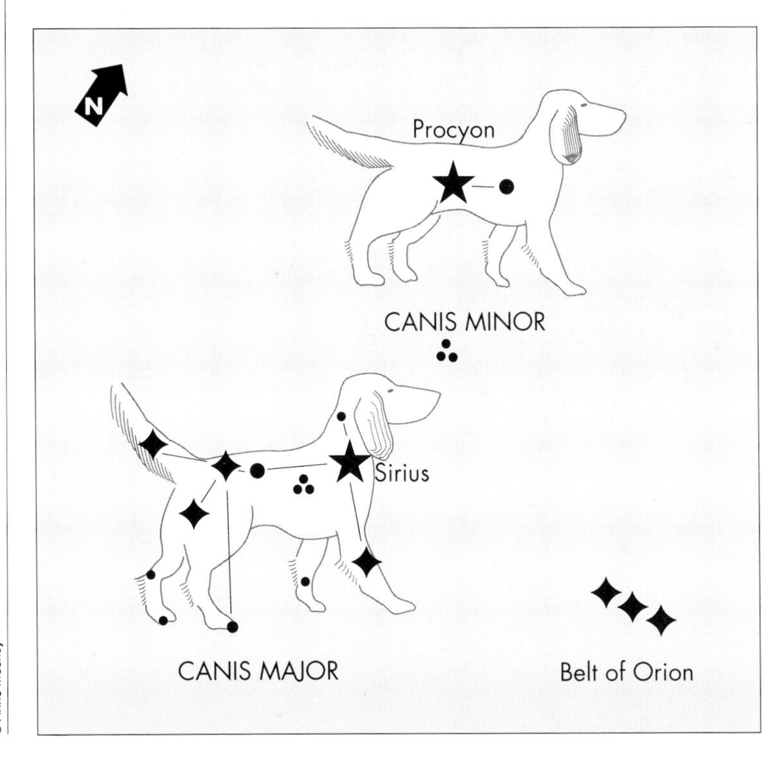
N
Procyon
CANIS MINOR
Sirius
CANIS MAJOR
Belt of Orion

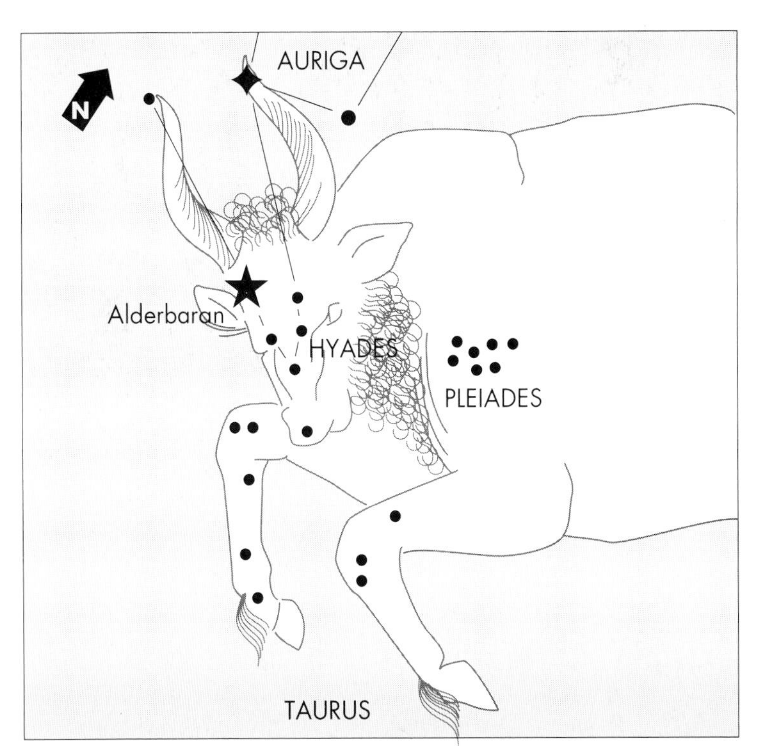
N
AURIGA
Alderbaran
HYADES
PLEIADES
TAURUS

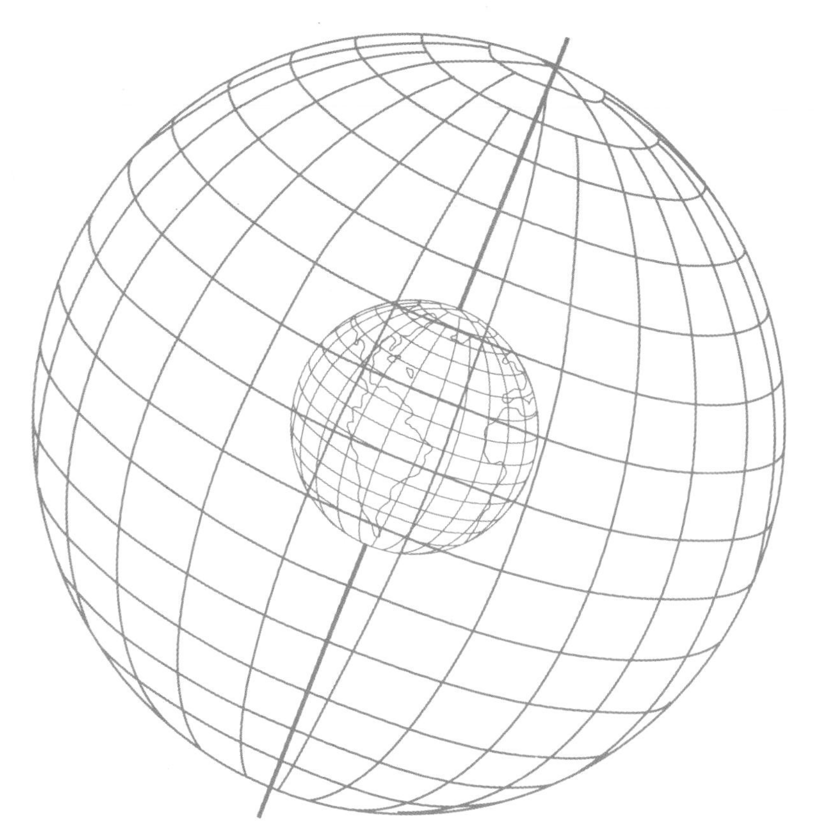

AT RIGHT is a representation of the celestial sphere. The stars are said to be at infinity (the outer grid), and the sphere is given lines of right ascension and declination that correspond to the Earth's lines of longitude and latitude, respectively. In the same way that Earth can be navigated by plotting a course along these arbitrary lines, so can the constellations be found each night by their right ascension and declination.

RIGHT ASCENSION AND DECLINATION

No matter where you are on Earth, the sky above you appears to be a dome covering Earth from one end of the horizon to the other. For practical purposes, think of yourself as being in a chair mounted to a wall at the center of a globe, looking out. You are only able to view one half of the globe from where you are sitting, so it appears to be a dome, just like the night sky.

Although the stars are not all an equal distance from Earth (as our globe representation would have it), for our purposes we can assume that all stars are the same distance away. This is because the stars are so far away from Earth that for an amateur astronomer's purposes, they can all be considered at the point of infinity. This imaginary distance at which all the stars are placed for observational reasons is called the *celestial sphere*.

Now imagine that this globe has lines of latitude and longitude printed on it, just like a globe of Earth. Astronomers use such a form of latitude and longitude, in conjunction with the boundaries of the constellations, to locate celestial objects and areas. To differentiate from Earth latitude and longitude, these measuring units have been given different names: The celestial equivalent of latitude is called *declination* (abbreviated Dec.), and the celestial equivalent of longitude is called *right ascension* (abbreviated R.A.).

All of the features of Earth latitude and longitude are maintained in this system of declination and right ascension. The North Pole of Earth coincides with the North Celestial Pole, and the Earth's equator is extended to the celestial sphere to form the celestial equator.

Declination is measured in degrees north or south of the celestial equator, with degrees north denoted by a plus sign (as in +11°) and degrees south denoted by a minus sign (as in –82.6°). The distance, of course, from the celestial equator to either pole is 90.° Because the universe is so large, even finding a particular degree is often not specific enough. Degrees of declination are therefore broken down further into sixty minutes, which are broken down further into sixty seconds each.

Right ascension is measured in hours, with the zero hour crossing through the westernmost point in the constellation

© Chip Isenhart/Tom Stack & Associates

This long-exposure photograph clearly shows the effect of the Earth's rotation. Since the stars (for all intents and purposes) are stationary, the streaks are caused by the movement of the Earth in relation to the stars.

Aries. It is measured along the celestial equator. Of course, one full traverse about the equator is equal to 24 hours (the celestial equator is, by definition, inextricably linked to the rotation of Earth). Therefore, one hour of right ascension is equal to 15°. Hours move from 0 to 24 eastward, so that a point 85° east of the zero hour is 4 hours and 40 minutes of right ascension, written R.A. 4′ 40.″

Often the location of stars or other celestial objects is given in right ascension and declination on star charts, particularly when discussing stars without using diagrams.

Time and the Speed of Light

One of the most interesting and least discussed facets of stargazing is the relationship of time and distance to an object being viewed. Few astronomers consider the role of time in the scheme of the cosmos, and certainly not while viewing a specific star. In fact, the question of *when* an object is being viewed didn't even come up until Einstein's theory of relativity.

Einstein showed that light travels at 186,000 miles (297,600 kilometers) per second, which seems very fast from a terrestrial perspective. And it is fast—a beam of light travels fast enough to circle the earth seven times in a single second. Yet the same beam of light, traveling from the nearest star to Earth—our sun, which is about 93,000,000 miles (148,000,000 kilometers) away—takes approximately eight minutes and thirty-five seconds to reach us. An observer on Earth, therefore, is viewing the sun as it was nearly nine minutes before, not as it is at that precise second.

The nearest star to Earth outside our own solar system is so far away that light from it takes 4.5 years to reach us; that star, [Proxima] Centauri, is 4.5 *light years* away from us. We are, therefore, seeing [Proxima] Centauri as it was four and a half years ago.

Multiply that by the scale of the universe as a whole and all sense of time and history goes right out the window. Scientists today are studying objects that are perhaps several billion light years away. That means it has taken the light from these objects several billion years to reach Earth, and since we can only see them as they were that long ago, they may not even be in existence today. Perhaps some celestial catastrophe may have occurred or is occurring now in the outer reaches of the universe, but Earth may not learn of it for several billion years.

Courtesy Celestron International

CELESTIAL TIME

One rotation of Earth relative to the sun is not the same as one rotation of Earth relative to the stars. Taking into account the effect of Earth's revolution about the sun, the difference is four minutes per day. Because this is so, astronomers must correlate Solar Time (one rotation of Earth relative to the sun, 24 hours) with Sidereal Time (one rotation of Earth relative to the stars, 23 hours, 56 minutes). It is this difference in time that affects the rising and setting of the stars by seasons. In fact, four minutes per day of difference means that the stars rise two hours earlier per month, or half an hour per week. That is why the monthly star charts published in astronomy magazines are accurate at a specific time early in the month, an hour earlier mid month, and two hours earlier at the beginning of the following month. It's easy to see why so many star charts are necessary during the year.

Astronomers keep time in Universal Time (abbreviated U.T.), which is the equivalent of Earth time at the Greenwich meridian. It is kept on a 24-hour scale with the zero hour occurring at midnight, Greenwich time. Universal Time can be figured by subtracting the number of hours west of Greenwich that observations are being made. For example, if an astronomer makes observations at 9:00 p.m. in New York, the Universal Time would be 16:00. Likewise, if observations are made at 3:37 p.m. in Los Angeles, the astronomer would give the time as U.T. 7:37. Universal Time is an important concept for astronomers, because the cataloguing of observations, if they are to be scientifically useful, must be comprehensible to other astronomers across the world. With the system of Universal Time, the dating and timing of observations is regularized throughout the world.

MAGNITUDE

Some stars shine brighter than others. This may be because one star is bigger than another, or because one star is closer to Earth than another, even though it is smaller or of similar size. This brightness, the brilliance of a star's light, is graded by its apparent magnitude from Earth; that is, how bright a person with average sight on Earth sees a star. (Astronomers sometimes grade stars by their ***actual magnitude***, which is the measure of what a star's brilliance would be if it were viewed in space from a distance of ten parsecs.) There are two magnitude scales in use today: The most popular was devised by Hipparchus more than 2,000 years ago, and the other, which also enjoys wide use, was invented by Johann Bayer in the 1600s.

Hipparchus's method measures magnitude of stars on a scale

The Pleiades, a naked eye or binocular group. Too spread out to be viewed with a telescope, the Pleiades are at their best in the camera, as a long exposure can bring out the faint nebulas in the region.

that originally ranged from 1 to 6, with stars of the first magnitude being the brightest and stars of the sixth magnitude the faintest. Obviously, this system only took into account the stars that were visible with the naked eye, since there were no telescopes or binoculars in Hipparchus's time. Eventually, the system was regularized and expanded to include much fainter stars, as well as define more accurately the brilliance of brighter stars. There are now measurable increments of brightness between each level of magnitude. Thus, Sirius, the brightest star in the sky, is actually magnitude −1.5 (negative numbers are used to denote stars brighter than magnitude 1; the sun is approximately magnitude −26). Polaris is magnitude 2, and the Hubble Space Telescope will be able to detect stars as faint as magnitude 27.

The second, less precise, method is Bayer's. It is based upon the constellations. In it, each of the twenty-four brightest stars in a constellation are denoted with Greek letters, so that the brightest star of the constellation is denoted by the letter *alpha*, the second brightest by *beta*, the third by *gamma*, and so on. Theoretically, any bright star could be named by its constellation and magnitude on the Greek letter scale. Unfortunately, errors in the naming of stars by magnitude were made. For example, the brightest star in Ursa Major is not Alpha Ursae Majoris but Epsilon Ursae Majoris, which should be the name for the fifth brightest star in the constellation. In fact, only three constellations of the eighty-eight recognized have had their five brightest stars named in the correct order. This is due to the poor quality of instruments available in Bayer's time, long before the development of the telescope. For this reason, the expanded method of Hipparchus is the one that is most widely in use today, and it is recommended for any serious astronomical purposes.

C H A P T E R   T H R E E

NAKED EYE ASTRONOMY

© Spencer Swanger/Tom Stack & Associates

Each of the lines in this long exposure photograph is a star or star cluster in the night sky. Photos such as these illustrate the concept of magnitude, with the brighter lines representing the stars with a greater magnitude.

NAKED EYE TRAINING

For many people, astronomy brings to mind images of scientists in white coats peering through the eyepieces of huge telescopes in immense observatories (perhaps while mumbling arcane formulas about relativity and neutrinos).

In fact, the most important observations a serious astronomer or casual stargazer will make require no telescope and no mathematical formulas. Just as a musician can't play Mozart until he or she has a firm grasp of musical scales, an astronomer can't begin to observe comets or stars until he or she understands how to look at the sky as a whole and break it down into smaller sections that are easier to remember.

It would obviously be impossible to memorize each star in the sky. However, learning the constellations is a relatively easy and valuable system for all stargazers to use, and there are many guides and systems available to help you.

One of the easiest, and perhaps the best, is that devised by H. A. Rey, an amateur astronomer and author of many books for children, including the *Curious George* series. Mr. Rey's method, which is at once useful for even advanced stargazing and simple enough for a child to understand, is perhaps the best known and most utilized of all modern stargazing memory aids. Although there have been many classical drawings of the figures represented by the constellations, they often were fanciful or elaborate drawings that were difficult to visualize in actual practice. Rey redrew the figures represented by each of the constellations, using only the actual stars as reference points. His illustrations of the constellations have been called dot-to-dot astronomy, but, though they are childlike, they are very effective memorization tools for both children and adults. For example, classical representations of Cetus, the Whale, usually show a large fish, often with all of the stars of the constellation well within the figure. Rey was able to trace a whale-shaped figure from the stars of Cetus. The amateur astronomer who looks at the stars of Cetus with Rey's drawing in hand will easily be able to associate a whale shape with that particular group of stars, and henceforth will see Cetus in that shape.

Another way of learning how to see the sky is simply to take a star chart for your area of the sky out to an observation point (a park, the roof of your house or apartment, or any other place that's relatively free of light pollution—the spillover of light from urban areas that is the bane of all astronomers) and memorize the constellations and patterns of stars it shows. If you have a couple of lawn chairs, ask a friend to join you; there is hardly a better way for an amateur stargazer to spend a few hours.

This naked-eye training should be at the top of the agenda for anyone even remotely interested in stargazing. The importance of learning to identify objects in the sky by sight alone can't be stressed enough. Many an amateur astronomer has looked up in the sky through binoculars or a telescope, found something that sparked his or her curiosity, and then spent hours going through astronomy maps, charts, and books trying to figure out just what it was that he or she saw.

If you are a beginning astronomer (or even if you have been stargazing for quite a while), take some time just to look up at the sky with a star chart in hand. Get to know the constellations, the brighter stars, the planets, and some of the other visible objects in the sky. The more you know about what you're looking at, the more enjoyable your time under the stars will be.

© Herman M. Heyn

Observation points such as this one at the top of Mt. Katahdin in Maine provide excellent views of the night sky, as they are elevated from any light pollution that may emanate from inhabited areas.

Hand Measurements

An easy way to measure distances in the night sky is by using your hands. Sailors and astronomers have been measuring between stars with this method for hundreds of years.

Stretch your arm out in front of you as though you are grasping for the sky. When you spread out your hand to its widest, from the tip of your thumb to the end of your pinky with your palm facing away from you, it covers about ten degrees of sky. Your index finger alone covers about one degree of sky. So if you know that a certain star is about 35 degrees to the left of, say, Mizar, and at the same altitude in the sky, you can find it by measuring three hand widths from Mizar plus three finger widths. While this is not a perfect system, nevertheless it is remarkably accurate for almost everyone. It seems that people with smaller hands tend to have shorter arms, and people with larger hands generally have longer arms, so for measuring purposes differences cancel each other out and two very different people are able to perform the same measurements.

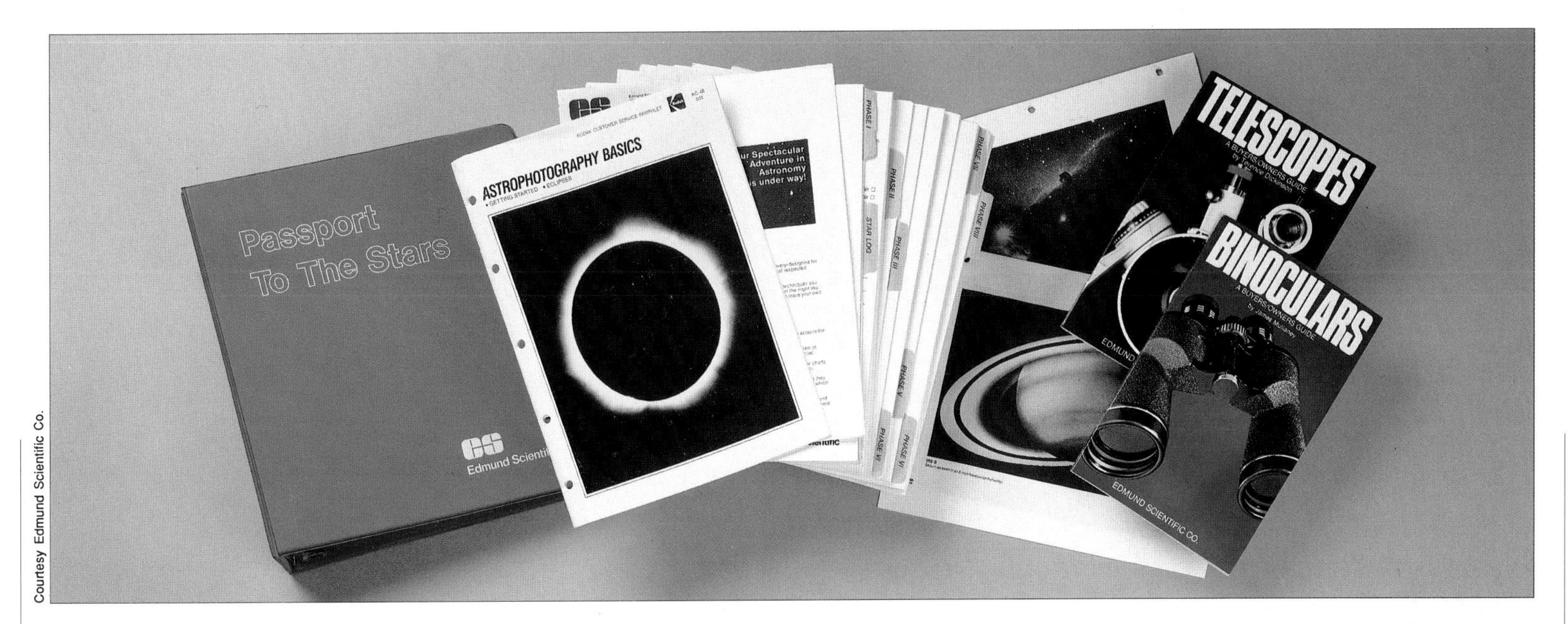

Courtesy Edmund Scientific Co.

Numerous references exist for the interested stargazer, from monthly astronomy magazines to very technical scholarly publications. However, no amount of study can compare with breathtaking views of the night sky such as the one AT LEFT.

LEARNING THE SKY

Learning to locate areas of the sky quickly and easily is important for two reasons: If a telescope or binoculars are to be used, the astronomer can easily find his or her way through the millions of stars; and it accustoms the eyes to "seeing" (a term astronomers use to describe the conditions and ability to view the stars) with telescopes. To really be able to employ a telescope, the eyes of an astronomer have to be trained to look at the night sky; it takes many stargazers several months to train their eyes to look efficiently through a telescope.

Each month, the two major astronomy magazines, *Astronomy* and *Sky and Telescope*, publish sky charts as well as lists of special objects or events that may be visible during that month. The charts in these magazines provide inexpensive, easy, and accurate maps of the stars. Another useful star chart is Wil Tirion's comprehensive *Sky Chart 2000.0*, which, while somewhat complex and detailed, is suitable for all stargazers.

Take your star chart with you to your favorite skywatching place as often as you can. Learn to see the sky with unaided eyes. As you become more knowledgeable about and familiar with the stars, you may want to search for harder-to-find celestial objects and events such as distant comets and nebulas. There are several books devoted to helping you to find just such objects and events. More specialized planispheres and more sophisticated sky charts are also available for helping you to find harder-to-locate objects. When you feel ready to search for the more challenging celestial objects, keep in mind that your ability to find objects by sight is paramount to your observations and will aid you no matter what your aim.

Using Red Light

While observing, you'll need a light source by which to read star charts or books or setting circles on a telescope, and so on. A penlight will do, but a red light is far better. Red light is cooler than white light and will not cause the same temporary retina burn that astronomers sit in darkened rooms to get rid of before observing. If you can find a flashlight with a red bulb, you've got what you need. If not, either make a light out of a red Christmas tree bulb or cover the face of your regular flashlight with red cellophane. Using red light, you'll be able to see in the dark without causing those imaginary stars that appear when you look at a light source too long.

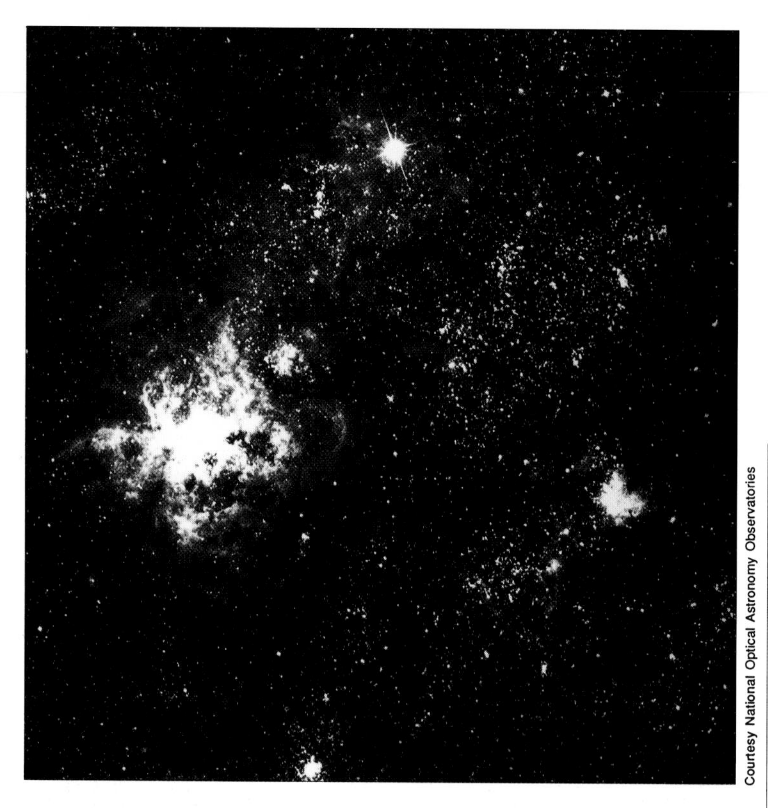

Courtesy National Optical Astronomy Observatories

The bright spot of illumination in the top third of this photo is Supernova 1987A. This photo was taken on February 26, 1987. Photographs of this area of the sky taken prior to the discovery of 1987A do not contain this object, which is still not completely understood.

WHAT TO LOOK FOR

The unaided eye can see celestial objects to approximately the sixth magnitude. That includes a lot of stars, the moon, asteroids, meteor showers, star clusters, and even the Milky Way galaxy. Depending upon where you are (since the light pollution of a city can diminish the visibility of the night sky), you may be able to see stars slightly fainter than sixth magnitude on nights of exceptionally good seeing.

Many cities have astronomy clubs or groups that offer sighting nights and star parties on evenings when specific sightings are possible, such as a comet.

MAKING OBSERVATIONS

One of the ways a stargazer can add enjoyment to sky watching is by recording observations. When you write down your observations, you can go back weeks, months, or years later and recall them. Another reason to record observations is for their scientific value. Perhaps you will witness a supernova one evening, or you will spend several days or weeks tracking a new comet you've just read about. These observations, particularly if sustained regularly over a period of time, may be useful to one of the many observing groups that may be studying that particular object or class of objects. Groups such as the Association of Amateur Variable Star Observers and the Association of Lunar and Planetary Observers require careful recording of observations, but even if you don't care to join one of these groups, observation records are an enjoyable way of cataloguing your sightings.

Observations should always include the following information:

1. Date and time of your observation: This is important, particularly if you may be comparing your observations with those of another astronomer. Time should always be noted in Universal Time.
2. Seeing conditions: Haziness, cloudiness, atmospheric disturbances, and so on, should be noted, as well as any light pollution from the lights of a city or the moon or sun (if you're making observations in twilight).
3. Optical instruments used: eyes, binoculars, or telescope. If optical instruments are used, lens aperture and magnification should be noted, along with any filters or other accessories used.
4. Observation: Be specific! List what was observed (if you don't know, list nearby reference points, such as, "Possible binary star in arm of Aquarius"), and give as many details as possible. Figurative descriptions and diagrams work well.

If you use these rules and guidelines, your observations are sure to be accurate, precise, and very enjoyable to look back on.

© Gary Ladd

The Antoniadi Scale

A simple way to describe seeing conditions for a particular night in your stargazing journal is to use the Antoniadi scale. This simple scale was developed by Eugene Antoniadi while he was making extensive and detailed maps of Mars. There are five steps in the Antoniadi scale:

I Perfect seeing, without a quiver.
II Slight undulations, with moments of calm lasting several seconds.
III Moderate seeing, with large tremors.
IV Poor seeing, with constant troublesome undulations.
V Very bad seeing, scarcely allowing the making of a rough sketch.

Once you get used to using the Antoniadi scale, it will take just a second at the end of each observation to describe your seeing conditions that evening.

© Gary Ladd

© Gary Ladd

These three views of Kitt Peak observatory show different seeing conditions. In fact, most stargazers find that excellent seeing conditions (I or II on the Antoniadi scale) occur less than ten nights each year.

© Gary Ladd

CHAPTER FOUR

STARGAZING WITH BINOCULARS

On some nights, skywatchers at Kitt Peak are treated to unexpected, but no less spectacular, sights.

Along with naked-eye stargazing, viewing the heavens through binoculars is one of the most enjoyable and relaxing methods of skywatching. The advantage of a good pair of binoculars is the increased light-gathering ability of the lenses, which may be able to capture up to several hundred times more light than the unaided eye.

Caution: Never look directly at the sun while using any optical instrument. Because telescopes and binoculars have the ability to gather many times more light than the human eye, they concentrate the rays of the sun into a beam of extremely strong light that can permanently damage your eyes. Be careful! Damage to your eyes is usually irreversible. There are special filters available from many telescope manufacturers for viewing the sun, and it is only through one of these filters that the sun should ever be viewed.

That said, binoculars are wonderful for seeing just about any object to the eighth magnitude. In fact, a pair of 7 × 50 binoculars will allow you to see the craters and seas of the moon, all of the planets except Pluto, many of the moons of the planets, and thousands more stars than are visible without optical aids.

An additional advantage of binoculars is that they can be used both for stargazing and nonastronomical purposes. In fact, many people already own opera glasses or binoculars for bird-watching, viewing sports, or other uses, and have never even considered using them for looking at the night sky. It may come as something of a surprise that some of the most exciting sights around have been right above them every night, and they've never even noticed.

OPTIC PRINCIPLES OF BINOCULARS (AND TELESCOPES)

Obviously, much more can be seen through a pair of binoculars or a telescope than with the naked eye. This is so because the lenses of an optical instrument function as a larger eye than the human eye. In fact, an optical instrument such as a pair of binoculars or a telescope really is just an extension of the observer's own eyes.

An optical instrument has two main functions: to gather light and to magnify images. It would seem at first that the most important of these is magnification: If any object were magnified enough, it should be viewable. This is a common, but incorrect, assumption.

The most important duty of any lens is gathering light. It is the amount of light gathered that determines how much can be seen. Put another way, if an instrument gathers more light, ever fainter objects can be seen, but if magnification of faint objects is attempted, a limit is reached and the image can no longer be magnified without serious blurring. It is for this reason that astronomers are said to be grasping for ever more aperture.

The aperture of a lens is merely its diameter. In binoculars, apertures of thirty to fifty millimeters are common, though it is possible to find binoculars of even greater aperture, particularly military field glasses. For the beginning stargazer, apertures of forty or fifty millimeters will allow plenty of new sights.

The other major function of optical instruments, magnification, depends upon the purpose. Binoculars commonly have

Courtesy Celestron International

Binoculars are especially good for viewing the sky when you want a wide field of view for seeing several objects or a very dispersed object, such as a nebula.

magnifications of 7 power or 8 power, written ×7 or ×8. For most binoculars, this is sufficient magnification, for although it is important to have sufficient enlargement of the image, there are other considerations. One such consideration is that there is a limit of usefulness to magnification, though this limit probably won't be met in a pair of binoculars.

Another is the field of view of the instrument. If the magnification is too high, the field of view becomes too narrow. For example, the moon covers approximately 1/2 a degree of sky when full. If the magnification of an optical instrument is too high, the field of view may be less than 1/2 a degree of sky, and you won't even be able to view the entire full moon.

Unlike a telescope, lightweight binoculars can be mounted on a camera tripod or held steadily in your hands. A good pair of binoculars should also be able to take some punishment, so be sure to purchase only binoculars with protective rubber around all glass surfaces.

TIPS ON BUYING BINOCULARS

Every stargazer, beginner or expert, should own a pair of binoculars for looking skyward at night. They are relatively inexpensive, useful, and don't demand nearly the degree of proficiency that telescopes do to get top results. Before you purchase a pair of binoculars, here are a few tips:

- *Aperture and magnification.* Decide first on the size and magnification you'll need for your purposes. Are you interested in viewing the moon, the planets, and other objects in our solar system? Aperture of forty to fifty millimeters and magnification of 7 or 8 power will be perfect. Are you interested mainly in observing one object or group of objects? A lot of aperture will get the fainter objects to appear, but more aperture means more money. A smaller pair of binoculars is more appropriate for sustained viewing over a longer period of time. (A new suburban binocular sport, Spot the Space Shuttle, calls for smaller binoculars.)

- *You get what you pay for.* This has always been true and will always be true, with few exceptions. The performance you get from a $30 pair of binoculars will not be the same as that of a $300 pair or even a $75 pair. At the same time, few people really have any need for a $300 pair of binoculars. A worthwhile pair, with decent optical quality, should cost under $100. Indeed, if you shop carefully, you can probably find a pair of good binoculars for closer to $50.

- *Buy sturdy.* Binoculars are famous for the way they slip out of even the most careful hands every now and then. The optics are fragile enough—don't purchase a pair of binoculars that seems flimsy. Look for binoculars with a hard rubber coating, a sturdy case, and, particularly, rubber around any glass surfaces. They should come with a good strong cord or strap to be worn over your neck as an added precaution.

- *Test the binoculars.* A dealer may or may not let you use them in a trial run overnight. If not, flip them up and focus in on the most distant object you can see through them. Do they feel firm in your hands? Are they comfortable, or are they so heavy that you can't hold them steady? Is there any distortion of the images? Even if you can't try them out, be sure you get an airtight guarantee that you can return them if there is a problem, and make sure you fill out the warranty forms.

© George East Courtesy Meade Instruments Co. (left and right inset)

© John Gerlach/Tom Stack & Associates

Binoculars are ideal for viewing near objects, like the moon.

HOW MUCH CAN BE SEEN WITH BINOCULARS

The range of celestial objects that can be seen with binoculars is often surprising to first-time users, especially city dwellers. Those who live away from the light pollution of the city, or those who can escape it easily, probably take their starry skies for granted. City dwellers, on the other hand, lose anywhere from five to forty percent of their night sky visibility because of city lights. However, one good thing about a pair of binoculars or a telescope is that they make the starlight brighter and the darkness darker at the same time.

Therefore, when a city-dwelling stargazer raises a pair of binoculars skyward, not only is the light from distant objects gathered more efficiently and magnified, but the optics provide much sharper contrast as well. This opens up a whole new world. Suddenly, stars appear out of nowhere, an occasional comet becomes visible, the sky is filled where it was mostly empty. Binoculars can improve a city dweller's stargazing immensely.

Meanwhile, the stargazer who lives away from the city lights will also see better with binoculars. While the city dweller may be able to pick up fourth- or fifth-magnitude stars with binoculars, the stargazer outside the city already can see these without optical aids. Binoculars improve seeing to the eighth magnitude; when conditions are optimal, even ninth-magnitude stars may be glimpsed.

There are many specific objects that are spectacular when viewed with telescopes, particularly the moon and the planets. A pair of binoculars with an aperture of fifty millimeters and a magnification of 7 or 8 power will partially resolve the rings of Saturn and at least three (perhaps four if conditions are good) of Jupiter's satellites.

C H A P T E R   F I V E

STARGAZING WITH A TELESCOPE

The famous Horsehead Nebula is one of the most interesting telescopic objects. The Horsehead Nebula is a "dark nebula", meaning its contents completely obscure all light that is behind it.

The four-inch (10-cm) refractor set-up above contains all of the elements for top-flight stargazing: a firm tripod, a balanced equatorial mount for tracking celestial movement, the eyepiece in position for optimal viewing (to reduce eyestrain and neck-strain), and, of course, a good quality telescope. The Trifid Nebula is an excellent object for a telescope with eight inches (20 cm) or more of aperture. AT RIGHT.

A st ... escope. Sooner ... urn to acquiring a telescope, though few really know what that entails. It's a good idea not to rush into such a purchase, however.

Nothing ruins a potential stargazer's enthusiasm quicker than the frustration of trying to learn to use a telescope before he or she is really ready. It isn't that using a telescope is difficult to learn, but several important skills must be mastered before a telescope can be employed with any usefulness. Too many novice stargazers rush out, buy the first piece of equipment they see, and are disappointed with the results.

Before you purchase a telescope, you should be very familiar with the night sky. Someone who can't locate at least half of the constellations and name the fifteen brightest stars in the sky should not be using a telescope. There is little to be gained by getting a telescope and just pointing it at things. Even if by chance you are able to view anything, you probably won't be able to tell anyone else what you've seen or repeat your observations on another night.

That said, it's probably useless to try to dissuade a determined stargazer from purchasing a telescope. In that case, forewarned is forearmed. The more knowledge the stargazer takes to the telescope store, the better chance he or she has of coming away with the telescope that will provide the most happiness and the fewest headaches for the money.

There are three types of telescopes: *refracting telescopes, reflecting telescopes,* and *catadioptric telescopes.* Each type of telescope has inherent advantages and disadvantages, as well as certain characteristics that may be either advantageous or disadvantageous depending upon the particular needs of each stargazer. In addition, before you purchase a telescope, there are many other factors to consider besides the telescope itself. Mountings and available accessories are just a few of these, and they will be discussed later in the book.

REFRACTING TELESCOPES

Refracting telescopes are so named because light that travels through them is bent, or refracted, by their objective lens. Refractors are the original optical instruments for astronomers: Galileo's telescopes were refractors. In fact, the refractors were the only telescopes available until Isaac Newton himself put together the first of a new type of telescope, the reflecting telescope, from John Hadley's designs.

In the refracting telescope, light is gathered by the objective lens, the large lens at the front of the telescope. The light waves are bent by the glass so that all of the light is projected to the smaller end of the telescope, which holds the eyepiece. The eyepiece is placed just behind the point of final focus of the image cast by the objective lens so that the image can be conveyed to the eye of the stargazer. Additionally, the eyepiece magnifies the

The principle upon which a refracting telescope works. The slightly convex objective refracts, or bends, the incoming light, in much the same way that the human eye sees. The image is focused at the end of the telescope tube and magnified by the ocular, or eyepiece.

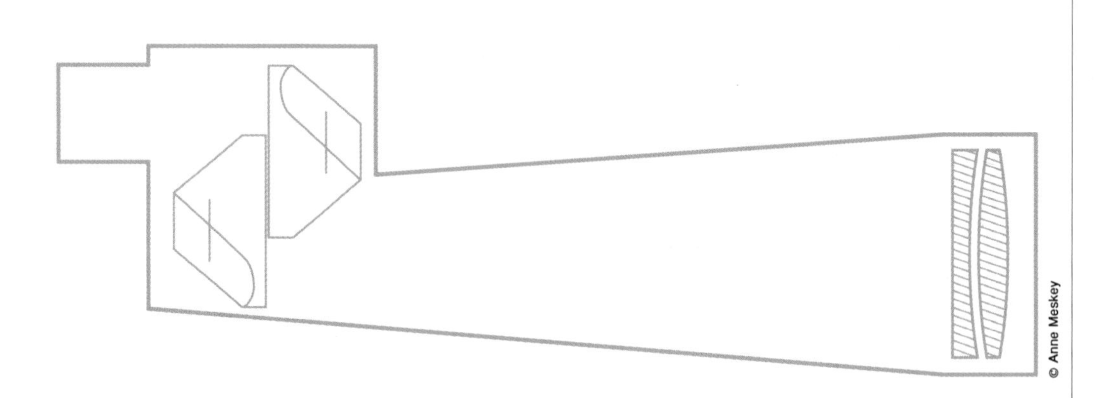

image (the degree of magnification depends upon the focal length of the particular eyepiece).

Refracting telescopes may be the best bet for beginning telescope users and for some advanced astronomers, particularly those looking for a "rich-field" (maximum field of view) telescope. Refractors are rugged and require very little maintenance. The housing (the telescope tube) is sealed on one end by the objective lens and on the other end by the eyepiece, which eliminates potential problems of dust and moving air currents inside the tube. At long focal lengths, refractors combine high-resolution capabilities with a wide field of view, for near-star comparison and for picking up larger areas of the sky.

The most obvious disadvantage of refracting telescopes is expense. Because a refractor relies on paraboloid lenses, which require precise grinding, optics for this type of telescope are costly. In fact, optics for a refractor cost about seven times more

Focal Ratio

An important consideration when buying a telescope is the focal ratio (f/ratio) of the optics. The focal ratio of a telescope is the focal length of the objective —the distance from the objective to the point of focus—divided by the aperture of the objective. For example, a reflecting telescope with a six-inch (fifteen-centimeter) mirror and a focal length of forty-eight inches (120 centimeters) has a focal ratio of 8, written f/8.

The longer the focal length of a telescope, the bigger the image. Therefore, for most types of stargazing, the higher focal ratios (f/10 and upward) are recommended. However, certain types of stargazing require a wider field of view with less emphasis on resolution, and for those kinds of viewing, focal ratios of f/6 and below are useful.

Before purchasing a telescope, you should decide whether your interests lie in looking at objects that require sharp resolution (long focal lengths) or a rich field of view (shorter focal lengths). Perhaps the best solution is a compromise —the middle range of focal ratios from f/6 to f/10 allow both types of viewing, to a lesser extent. Before buying, discuss focal ratio with an experienced stargazer or a knowledgeable telescope salesperson.

Courtesy Celestron International

The Sombrero Galaxy, like our Milky Way Galaxy, is an excellent example of a spherical galaxy. FOLLOWING PAGE, *dusk is often the best time for stargazing, though you must be careful if the sun is still above the horizon not to let it near your field of view.*

than optics for a reflecting telescope. It is therefore an unfortunate truism that each inch of refractor aperture costs about twice as much as each inch of aperture on a reflector or catadioptric telescope. Put another way, a four-inch refractor costs about the same as an eight-inch reflector or catadioptric.

Another disadvantage is the size of a refractor. To cure a small but unavoidable defect of a refracting lens called chromatic aberration, the focal ratio must, as a rule, be at least three times the aperture of the lens. For example, a four-inch (ten-centimeter) refractor should have a minimum focal ratio of f/12. Thus, the focal length of the telescope would have to be a minimum of forty-eight inches (120 centimeters), or four feet. Not surprisingly, a four-foot-(1.2-meter-) long telescope is both unwieldy to transport and difficult to mount vibration free. Also, because of the length of the telescope tube and the location of the eyepiece on the bottom end of the tube, a relatively tall mount is necessary, which is usually both expensive and prone to vibration. All these factors make refracting telescopes with more than about four inches of aperture impractical for the amateur astronomer.

Still, among telescopes with smaller apertures, the refractor is unbeatable in all respects but price. Images are clearer; the tube remains clean and clear and requires little maintenance; and mobility is not such a problem when the aperture, and therefore the focal length, is smaller. However, if apertures of greater than about four inches (ten centimeters) is desired, refracting telescopes are prohibitively expensive and difficult to use. For this reason, few refractors with objective lenses larger than four inches (ten centimeters) in diameter are manufactured today.

© John Sanford/Photo Network

The catadioptric telescope shown here employs features of both the Schmidt camera and the Cassegrain reflector to reduce the size of the tube needed to reflect the image of the stars.

REFLECTING TELESCOPES

In a reflecting telescope, an objective mirror at the bottom of the telescope tube, called the primary, replaces the refractor's objective lens. The tube is open at the top to let incoming light pass down to the bottom of the tube. Light waves travel down to the primary mirror, which collects all the incoming light and reflects it back up the tube to a small, flat mirror placed at an angle in the center of the tube. The images cast from the primary mirror are deflected sideways by this small mirror called the flat. An eyepiece is placed near the top of the telescope tube perpendicular to the incoming light, and the images are deflected by the flat to the eyepiece.

This design, the Newtonian reflector, was invented by John Hadley in England in the mid-seventeenth century. It takes its name from the first reflecting telescopes, which were built by Isaac Newton himself from Hadley's designs. The Newtonian reflector quickly caught on with astronomers and has been the telescope of choice for the majority of stargazers, both amateur and professional, for at least the past half century. A recent development in telescopes, the catadioptric, is beginning to replace the Newtonian reflector in popularity, though both reflectors and refractors still have a considerable following.

The greatest advantage of the reflecting telescope is its cost per inch of aperture. The cost-to-size ratio is the primary difference between reflectors and refractors. This is because there is little use for very small reflecting telescopes—any savings of small mirrors over small lenses is minimal—while at the same time it is much easier to make a large mirror than a large objective lens. This is particularly true for amateur stargazers, who don't receive government grants and foundation appointments to offset spiraling costs of technology.

The Newtonian reflector provides other advantages in addition to its low cost. Since it uses a mirror instead of a lens, light waves are not bent and therefore chromatic aberration is not present. Thus, the Focal-ratio-equal-to-three-times-aperture rule does not apply. Reflectors can have shorter focal lengths than refractors, so that an eight-inch (twenty-centimeter) reflector can have an f/ratio of f/4.5 and thus be just over three feet (a meter) long. An eight-inch (twenty-centimeter) refractor would have to be sixteen feet (4.8 meters) long to correct for chromatic aberration; it's easy to see the advantage of mirrors in larger apertures.

Another advantage of reflectors is that only two optical surfaces (the faces of the objective mirror and the flat) are employed, lessening the potential for optical flaws that can significantly reduce the contrast, definition, and overall quality of the image. Also, since the eyepiece is near the top of the telescope tube, the Newtonian reflector can be mounted relatively low to the

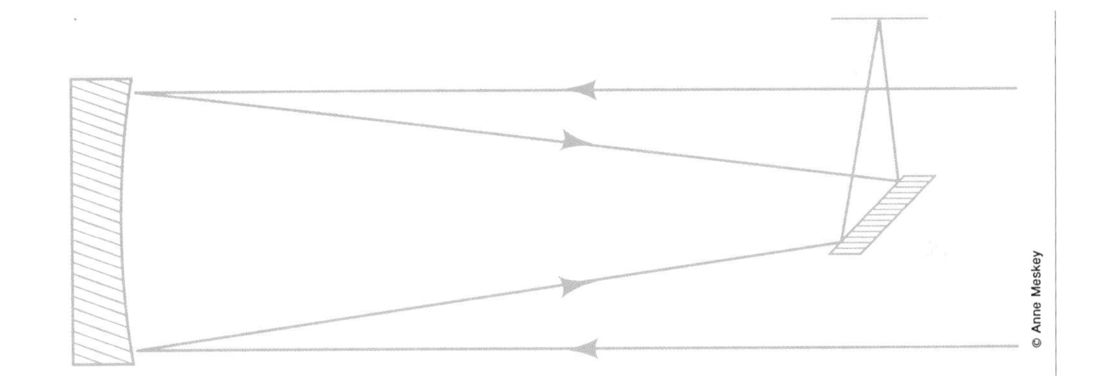

In a reflecting telescope, light enters the telescope tube and travels to the base, where a curved mirror, the primary, is mounted. The light is reflected from the primary to the smaller flat, which then directs the light through a hole in the tube where the ocular is placed to magnify the image.

ground, which means a lower center of gravity, easier and sturdier mounting, and a decrease in potential strain to the back and neck of the stargazer.

Though the Newtonian reflector has several advantages over a refractor, it also has some characteristics that can be disadvantageous. Because of the relative imprecision of mirrors as compared to ground-glass lenses, definition suffers near the edges of the field of view in reflectors, particularly at low power. Also, because of its placement at the center of the tube, the flat blocks a small portion of the incoming light, causing slight degradation of the image, particularly at shorter focal lengths.

Another major disadvantage to reflectors is the amount of maintenance necessary to keep them in top working condition. While this maintenance is not extensive, it is much more than is required for a refractor. For example, because the telescope tube is open (it must be—any covering, even a thin transparent film, would reduce the amount of light entering the telescope by appreciable amounts), the problem of dust on the optics is always present. Also, the mirrors of a reflector fall out of alignment much more quickly than the optics of a refractor; the owner of a Newtonian reflector may have to realign the mirrors each time the telescope is used in extreme cases. Realigning mirrors is a simple but boring and time-consuming task, much like tuning a musical instrument. Additionally, the mirror coating will need to be resilvered every few years (the process is still called silvering, even though mirror coatings are now aluminum or an aluminum alloy). The coating tarnishes over time, which cuts down on its reflectivity and reduces the definition and contrast of the image. To allay tarnishing as much as possible, covers must be placed on both mirrors at all times when the telescope is not in use.

Another type of reflecting telescope developed just after the Newtonian reflector is the Cassegrain reflector, which works on an efficient principle to elongate the focal length of a telescope. Like the Newtonian reflector, the Cassegrain reflector is an open tube with a mirror at the bottom end. However, in the Cassegrain reflector the primary mirror has a hole in its center large

Courtesy Celestron International

The Omega Nebula, ABOVE, *is a deep sky object that requires a fairly large telescope to view. However, thousands of other objects are visible with a smaller reflecting telescope like the one pictured* AT RIGHT. FOLLOWING PAGE, *the Lagoon and Trifid Nebulas, both visible here, are emission nebulas, clouds of hydrogen gas excited into luminescense by nearby stars.*

enough to hold an eyepiece. Light is reflected from the primary mirror to a curved secondary mirror, which is placed where the flat would be in a Newtonian reflector. This small mirror reflects the light directly back at the hole in the center of the primary mirror to the eyepiece. In this way, the portion of the primary mirror that would receive almost no original light because of blockage from the secondary mirror is removed in favor of an eyepiece. The secondary mirror throws the image back the same distance it has traveled, increasing the focal length by twice as much or more in the tube. In this way, a long focal length can be combined with a large mirror in a relatively short and easy-to-manage telescope.

Surprisingly, this type of reflecting telescope is very rare in amateur use. However, the principle of the Cassegrain is widely used in amateur-level catadioptric telescopes, which combine the use of a mirror and a lens.

The Limits of Viewing

© Greg Vaughn/Tom Stack & Associates

What you can see depends upon what you are using to see with. As the table below shows, aperture is the most important factor in seeing faint stars. Here, then, is a list of the limits of viewing:

Aperture	Faintest Visible Magnitude	Separation (Arc-seconds)
Human eye	5–6	8″
50 mm binoculars	8–9	3–4″
3-inch (7.5-cm) telescope	11–12	2″.5–3″
6-inch (15-cm) telescope	12–13	1″–1″.5
8-inch (20-cm) telescope	13–14	0″.5
40-inch (100-cm) Yerkes refractors	18	0″.11
Hubble Space Telescope	27	0″.02

This list assumes optimal viewing conditions. However, if there is significant light pollution or air turbulence, the faintest visible magnitudes may be appreciably higher. If, for example, you are viewing from your backyard in downtown Omaha, your eight-inch (twenty-centimeter) telescope may only show celestial objects to the tenth or eleventh magnitude, while the same telescope in Central Park in New York City may show celestial objects only to the eighth or ninth magnitude.

Courtesy Celestron International

The Pleiades, as a group, are best viewed in small telescopes or binoculars. Larger telescopes and long exposure photographs can bring some of the nebulous material around each of the stars in the cluster into view.

CATADIOPTRIC TELESCOPES

The real newcomers to the world of astronomy, catadioptric telescopes have only been available commercially since the early 1970s, and thus their virtues and faults are still widely debated among users. The catadioptric that is most commonly available and most widely used is the Schmidt-Cassegrain telescope, sometimes called the SCT. The Schmidt-Cassegrain is a combination of the Schmidt camera and Cassegrain-style optics. The Schmidt camera, which was developed in the 1940s, uses a ground-glass lens-type plate to correct the image cast by a mirror and was originally developed to take high-quality astronomical photographs. The Cassegrain mirror system is a method of elongating the focal length that is ideally suited to the Schmidt camera.

In the Schmidt-Cassegrain telescope, light travels through the corrector plate to the primary mirror. The light is reflected back to the secondary mirror, which is mounted just behind the corrector plate. The image is then reflected back through the hole in the primary mirror to the eyepiece, making the light travel a long distance (longer focal length) in a short housing (resulting in telescope tubes of twenty-four inches or less).

There is no optical advantage to a Schmidt-Cassegrain telescope over a reflecting telescope. In fact, the secondary mirror of a catadioptric telescope causes more obstruction to the image than the flat of a reflector, so the image definition and contrast can be slightly worse in the catadioptric. The difference is minimal, however. The catadioptric's main advantage is that its optics allow it to be considerably smaller than a reflector of similar aperture. Not only is a Schmidt-Cassegrain easier to transport and set up, but it is also easier to mount firmly because of its compact size. This translates to easier access to larger apertures for the amateur stargazer. Another advantage of a catadioptric telescope is that the corrector plate covers the upper end of the tube, thereby making the design a closed-tube assembly, which eliminates dust in the housing and other maintenance problems.

Catadioptric telescopes have several disadvantages in addition

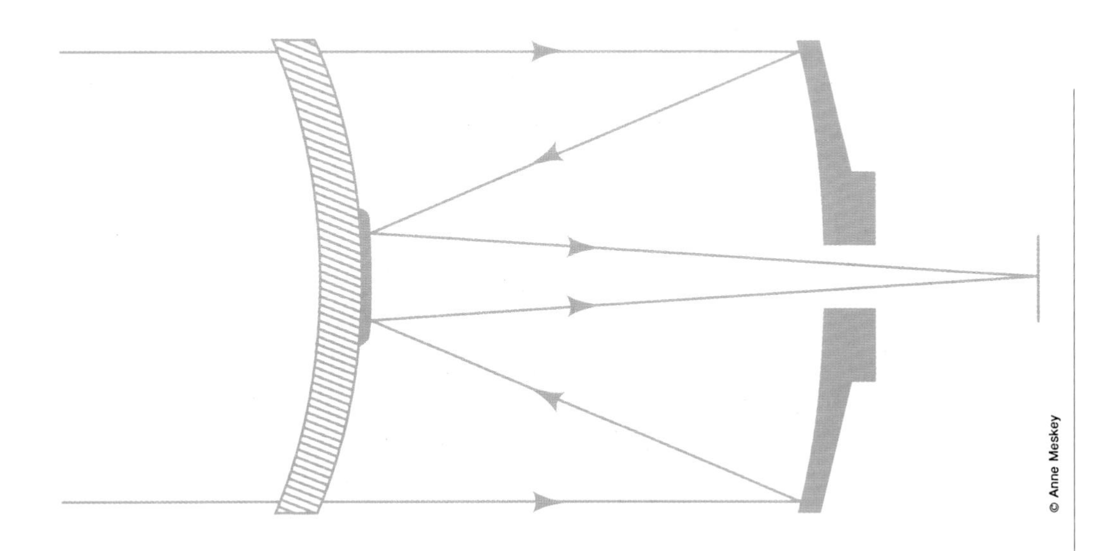

In the Schmidt-Cassegrain system, light passes through the corrector plate and is reflected from the primary mirror at the base of the tube to the secondary mirror, which is mounted to the corrector plate. The secondary then reflects the image back through a hole in the primary to the ocular. The back of the secondary blocks a small portion of light to the primary; this blocked area is where the ocular is placed, thus taking advantage of unused mirror area.

to the unavoidable built-in obstruction of the secondary mirror. One is that incoming light strikes several optical surfaces before it reaches final focus, thus increasing the chance that somewhere along the path of the light inside the telescope tube an imperfection will play havoc with the image. With care, however, this problem can be avoided. A catadioptric telescope, with its many optical surfaces, is more difficult to realign than either a reflector or a refractor, though it is certainly not an enormous undertaking to learn. The final disadvantage of a catadioptric is cost: The Schmidt-Cassegrain design is more expensive than a Newtonian reflector of the same aperture, and the other popular catadioptric design, the Maksutov-Cassegrain, is still more so. However, both of these catadioptric telescopes are far less expensive than a refractor of similar aperture. Thus, the portability and ease of use of the catadioptric styles outweigh their defects in the minds of many amateur stargazers.

Chromatic Aberration

Chromatic aberration is a problem inherent in any lens-based optical system. When light hits a lens, it is separated into its spectrum of colors. In a perfect world, a second, opposing lens could recombine the spectrum exactly into its original form of light. There is, of course, no perfect lens pair that can recombine a light spectrum exactly. The light can for the most part be reconstructed, but the few star light waves of the spectrum, the *chromatic aberration*, may distort the image or cause colored rings of light to surround the image.

There are two methods of eradicating chromatic aberration: Choose a refracting telescope with a long focal length, which compensates for this undesirable effect, or choose a reflecting telescope, which has no lens and therefore no chromatic aberration.

Courtesy TeleVue

A four-inch (10-cm) refractor on an equatorial mount and a three-inch (7.5-cm) refractor on an altazimuth mount. For tracking celestial objects, the equatorial, with its setting circles (the white ruled bands at the bottom of each axis), is much easier to use.

MOUNTINGS

Just as important as the size and type of telescope is the mounting on which the telescope will be placed. It is easy to see why: Suppose you spend all of your money on a big, beautiful Newtonian reflector, and you plan to mount it on your camera tripod. A mount such as this would ruin your love of stargazing instantly.

A mount such as a camera tripod is not nearly sturdy enough to hold any but the most lightweight telescopes without allowing excessive vibration each time you approach the eyepiece. There is truly nothing more annoying to the stargazer than a telescope that jiggles when you use it. In fact, the most infuriating mounts are those that are almost, but not quite, sturdy enough, so that you can get the image in decent focus in the eyepiece but it continually flickers and jumps just enough so that it cannot be resolved clearly. To avoid this frustrating situation, and the headache that accompanies it, investing in a good, large, heavy, sturdy mount is strongly recommended.

There are two main types of mounts for telescopes: altazimuth and equatorial. A standard camera mount is an altazimuth-type mount—it allows the camera (or telescope) to be moved to any altitude and along any azimuth, or bearing (that is, it swings both horizontally and vertically). The camera (or telescope) rotates around an axis perpendicular to the earth, pointing to the observer's zenith. An equatorial mount is actually a tilted altazimuth mounting. The telescope is mounted to an altazimuth-style mounting that rotates, when properly aligned, around the celestial north pole. In effect, the celestial north pole is the zenith of the telescope mount in an equatorial mounting. Because the telescope swivels around the same axis that the earth rotates around, an equatorially mounted telescope can offset the effect of Earth's rotation and thereby allow the stargazer to follow the stars in their paths across the night sky.

The effective result of the altazimuth mount is that the stargazer's visible horizon is the equator for the telescope. Conversely, the equatorial mount allows a telescope to be positioned along the lines of declination and right ascension, with the celestial equator as the equator for the telescope's field of rotation. This alignment with celestial declination and right ascension also makes star finding very simple. Most equatorial mounts come with dials imprinted around their movable axes called setting circles, one denoting right ascension and the other denoting declination. This makes finding stars by their positions of right ascension and declination even easier.

An equatorially mounted telescope tracks movements of celestial objects much more easily than an altazimuth-mounted telescope. With an equatorial mount, when the object to be viewed is located, the declination axis can be locked onto its position.

The Dobsonian Mount

A fairly recent development in mountings that has become popular for larger telescopes, particularly those that require mobility, is the Dobsonian mount. A large yet portable version of the altazimuth mount, the Dobsonian has been used with great success by many amateur stargazers. It consists of a revolving platform upon which two struts are affixed. The telescope is mounted at the top between the struts, and it can be rotated to any altitude and revolved to any position in the sky. Though big and unwieldy, the Dobsonian mount is one of the few truly sturdy portable mount designs, and it has the advantage of being fairly easy for the amateur to make at home; or, if putting together a mount at home is impractical, it is relatively inexpensive to purchase.

The Dobsonian mount was developed by John Dobson, the leader of an observing group known as the San Francisco Sidewalk Astronomers. Dobson and the Sidewalk Astronomers are devoted to exposing as many people as possible to the wonders of astronomy. Dobson is famous for holding impromptu star parties in "heavy-traffic" areas to attract as many people as possible—hence the name Sidewalk Astronomers. He is also well known for his ability to make and acquire telescopes, even on very limited funds. In this respect, Dobson is something of a Johnny Appleseed of telescopes, because he arranges for all of his telescopes to be donated to observing groups, schools, and other organizations that may wish to let the general public use them, both in the United States and abroad. According to Grant Fjermedal in *New Horizons in Amateur Astronomy*, Dobson himself has had more than one million people peer through his telescopes. He is perhaps responsible for enlisting more amateur stargazers into astronomy than any other single person in the world.

The Andromeda Galaxy, a spiral galaxy, is one of the nearest galaxies and therefore is a favorite object of many astronomers.

With only the right ascension axis movable, the telescope can be maneuvered manually or with a clock drive, an electrical gear-driven motor drive, along the axis of right ascension, which is the axis along which the stars travel. Unfortunately, the altazimuth mounting is unable to perform nearly so efficiently. Both the altitude and altazimuth axes must be manipulated to track a star, and a clock drive cannot be used with an altazimuth mounting. This makes guided astrophotography impossible with an altazimuth-mounted telescope.

However, the equatorial mounting is not without its faults. Since it is a modified altazimuth style, in which the mount itself is displaced from the stand upon which it rests (it is nearly impossible to tilt the whole stand to align with the celestial north pole from most places on Earth), the telescope and mount must be counter-balanced to even out the weight distribution on the stand. This sometimes causes clumsy movement of the telescope and can make transportation much more difficult because of the additional weight. Additionally, the equatorially mounted telescope is often difficult to maneuver when pointed near the horizon, and adjustments in positioning are often necessary when viewing celestial features that are not high in the sky. The final disadvantage to an equatorial mount is cost: An equatorial mount can cost up to twice as much as a standard altazimuth mount.

Other disadvantages aside, the equatorial mounting is probably worth the additional cost. This is particularly true if you are interested in astrophotography, because long exposures would be impossible with an altazimuth-mounted telescope and camera. In fact, there are very few cases where the altazimuth mounting is preferable to the equatorial. However, if you can afford either a stable altazimuth or a less stable equatorial mount, always go with the more stable mounting. Though star-tracking across the sky will be considerably easier with the equatorial mount, it makes no sense to be able to track the motions of stars and planets but not be able to get them into focus because of an unsteady mount.

OTHER CONSIDERATIONS

There are several secondary considerations for the potential telescope purchaser.

Eyepieces, also called oculars, have two functions: to magnify the image cast by the objective and to focus the image for the observer. Eyepieces vary greatly in degree of magnification, price, and quality. Most telescope owners require three types of eyepieces: a low-power eyepiece for wide-field viewing, a high-power eyepiece for taking closer looks at interesting objects, and a middle-power eyepiece to combine these two factors. A good general rule for choosing magnification for each eyepiece is the 10-25-50 rule. The low-power eyepiece should have a magnification of about ten times the aperture of the telescope in inches, the medium-power eyepiece should magnify about twenty-five times the aperture in inches, and the high power should magnify about fifty times the aperture in inches.

Eyepieces come in a wide variety of configurations. The most popular are the orthoscopic and plossl types of eyepieces. These have the advantage of giving good performance with little eyestrain. Other eyepieces, most notably the Ramsden and Tolles types, can perform better than the orthoscopic and plossl eyepieces, but they can also cause severe eyestrain and are thus only useful to professional astronomers.

One other type of eyepiece is important to mention. That is the Barlow lens, a "negative" lens because it diverges rather than focuses incoming light. For that reason, the Barlow can never be used alone. However, if a Barlow lens, which is also known as a Negative Amplifier lens, is placed in front of the focal point of the telescope, it will double or even quadruple the magnification

Courtesy TeleVue

Be certain when purchasing a telescope that a variety of compatible eyepieces and other accessories is available.

Courtesy TeleVue

A wide range of eyepieces is available to the stargazer. When you purchase a telescope, you should confirm with the salesperson that all types of eyepieces are compatible with your new telescope.

of a standard eyepiece by effectively increasing the focal length of the telescope tube.

If the telescope you are considering purchasing is a refractor or a catadioptric, then a star diagonal will probably be necessary. A star diagonal is an L-shaped attachment with a slanted mirror inside. When it is placed in the opening where the eyepiece would be attached, it allows the eyepiece to be attached to it, perpendicular to the telescope tube. This is a much more comfortable position for viewing, particularly if the telescope is on a mount under six feet (1.8 meters) tall. Obviously, reflectors don't require star diagonals because the secondary mirror of a reflector performs the same function.

Another important consideration in buying a telescope is the finderscope that comes with it. A finderscope is a small telescope that is attached to the main instrument. It is usually of magnifica-

The Magnification of an Eyepiece

One function of the eyepiece is to magnify the image being transmitted by the objective. However, when you purchase an eyepiece, you don't purchase it by its degree of magnification, but rather by its focal length. This is because the degree of magnification of an eyepiece is dependent upon the focal length of the telescope with which it is being used. In fact, the magnification of an eyepiece is the focal length of the telescope divided by the focal length of the eyepiece.

It's easy to understand why this is so: Suppose your telescope comes with an eyepiece that has a focal length of 25 mm. If the focal length of your telescope is 900 mm, the eyepiece will magnify the image thirty-six times, or ×36. If you put the same eyepiece in a telescope that has a focal length of 1000 mm, the image will be magnified by ×40. And if you put the eyepiece in a catadioptric telescope with an effective focal length of 2000 mm, the eyepiece will magnify the image ×80. Likewise, if you place an eyepiece with a focal length of 4 mm in that catadioptric telescope, the eyepiece will magnify ×250. This is because the image travels the equivalent of 2000 mm from the objective to its focal point, but only 4 mm from its focal point to the lens of the eyepiece, which is 1/250 of the distance.

A point to keep in mind with magnification is that there are limits of useful magnification. If you wish, you can magnify an image a million times, but it will be a murky, unrecognizable mess. The limit of an eyepiece's resolution, called the Dawes limit, is about sixty times aperture in inches. Therefore, an image cast by a three-inch (7.5-cm) refractor shouldn't be magnified more than about ×180, while the image cast by a twelve-inch (30-cm) reflector can be magnified up to ×720. However, there are very few times when this degree of magnification will ever be necessary or useful; in fact, even in the largest amateur telescopes, useful magnifications of greater than ×300 are uncommon.

tions between $\times 5$ and $\times 10$ and typically has apertures of between one and two inches. The finderscope is aligned to the main telescope so that it can be used to locate objects, much in the way a scope is mounted on a gun to aid in aiming. The smaller finderscope encompasses a much wider field of view than the main instrument, which makes finding specific stars or objects much easier by "star hopping"—jumping from star to star in a specific area or constellation until the particular object or area to be viewed is located.

An accessory that is usually necessary for observers at one time of the year or another is a dew cap. A dew cap is an attachment, usually just another tube, that fits onto the end of the telescope to prevent dew from forming on the objective. Most telescope manufacturers sell dew caps, some of which are even heated electrically (and therefore have to be plugged into a power source). However, a dew cap can be made by extending any material the length of the telescope tube. A funnel shape is better than just a straight tube, as a straight tube that extends out too far can reduce the effective aperture of the objective. Another dewing problem is a film forming on the eyepiece. This is easily avoided by wrapping the eyepieces in cloth and carrying them in your pockets for a few minutes before they are used. The warmth of body heat will keep them dew free for quite some time, and when a dew film does begin to form, the eyepieces can be wrapped in cloth again and put back into your pockets for a few more minutes.

A final consideration for a telescope buyer is whether to build a dedicated observatory. There are several advantages in having a permanent, useful home for your telescope, particularly if it is large. The main advantage is that a dedicated observatory provides a place to keep a telescope mounted at all times, thereby allowing it to be mounted on a permanent axis that has been cemented into the ground—the sturdiest, most stable mount possible. An observatory can be heated and remains dry. Also, an observatory can be placed in a strategic area—a hill or in a dark field, for example—and remain in place indefinitely. A final advantage is that an observatory can provide a storage place for all astronomical equipment, from eyepieces to the main instrument itself (and if you've ever tried to find a good place to keep a five-foot-long, eighty-pound (1.5-m, 36-kg) telescope, you understand the value of the right storage space).

Of course, the main disadvantage to building your own small observatory is its cost. Depending upon the size, location, and design of the observatory, it can cost several thousand dollars to erect. Also, an observatory has specific (and somewhat difficult) design requirements: It must be big enough for a telescope and observer to move comfortably within, and it must have a roof that is easily removable as well as sealable from rain, wind, and dirt. And, because of its size requirements, an observatory requires a small chunk of real estate upon which conditions for astronomical observation are optimal, and such a plot may be impractical, unavailable, or just plain unaffordable for most people.

That said, for the committed stargazer there is no substitute for having your own small observatory. There are several companies that offer kits to show you how to build your own observatory, and there are even a few companies that specialize in building astronomical observatories in backyards across America. Whether you want to do it yourself or pay to have one built, a personal observatory can enrich your observing experiences immensely.

This beautiful four-inch (10-cm) refractor, with a great selection of eyepieces, an equatorial mount, a 35 mm camera and adaptor, and a good star chart, are all you need to immerse yourself in excellent stargazing and high-quality astrophotography.

PURCHASING A TELESCOPE

There is no substitute for observing experience in helping you choose a telescope. Since getting experience in telescope observing when you don't own a telescope is difficult, the next best thing is to borrow someone else's experience. If there is a stargazing club in your area, contact it and get involved. Such clubs are always looking for people who share their enthusiasm for astronomy. If you let some of the members know that you'd like advice on buying a telescope, you'll be inundated with telescope tips, basics, and stories.

Stargazing groups often have star parties or observation nights open to the public. Attending such an event will provide a chance to peer through several telescopes, which will give you both observing experience and a basis for comparison of telescopes. This is particularly important for those who have never actually used a telescope.

Some people think that when they look through a big telescope, suddenly all the stars in the sky will look as big as the moon. They are sure to be disappointed. Any telescope will show you many more objects in the sky than you can see with just your eyes. But no telescope will ever make any star (except our own sun) look any bigger than a very small point of light. Even the nearest stars are much too far away to be resolved as objects, even by the largest telescope on Earth.

A telescope will show you thousands more stars than your unaided eyes can see, plus many other celestial objects that are invisible to the naked eye. A telescope can give excellent views of the objects in our solar system—everything from Jupiter's moons to the minor planets (asteroids) to man-made satellites. Most importantly, a telescope will allow you to see the activity of the heavens—albeit as a distant viewer—and reveal some of the true natural wonders of the universe.

Once you've got a firm grasp of what you'll be looking for and what kind of instrument you want to use to see what you're looking for, you're probably ready to go to the store with checkbook or credit card in hand. Here are a few additional things to consider and to discuss with a salesperson:

Courtesy Celestron International

It's easy to see how a spiral galaxy like this one could be confused with a nebula. However, one good clue that this is a galaxy is the spiraling "arms" that curl out from the center of the object.

• *What accessories are included in the price of the telescope?* How many eyepieces come with the telescope? Does it have a carrying case? What kind of finderscope is provided? What type of mount? (You should test the mount's stability before you buy it.) Does the telescope come with a motorized clock drive? If so, what is the drive's power source? How are the optics protected?

• *What accessories are available for this telescope?* Are a full range of eyepieces available? Filters? Can the telescope be computer driven? Can it be adapted for astrophotography?

• *What kind of warranties are provided by the manufacturer and seller?* Does the store have a return policy? (There should always be a trial period.) In case of problems, what kind of service is available? How and where is it found?

• *When will the telescope be delivered?* Is the telescope in stock? If not, how long will it take to receive? (Get a firm commitment, make the payment contingent upon prompt receipt of the telescope on the stated delivery date, and don't be surprised if the salesperson says the telescope can't be shipped from the manufacturer for a year or more—delays of this type are not uncommon for some brands of telescopes. If you don't want to wait, you may need to find a store that has the model you want in stock or else choose a different model that is in stock or has a short delivery time.)

If you keep in mind your specific stargazing situation—what kinds of celestial objects you'll be looking for and where you'll be doing your looking—you should be able to make a good choice. Then all you have to do is learn to use your new telescope correctly.

HOW TO LOOK THROUGH A TELESCOPE

Looking through a telescope seems simple—just point and shoot like a camera, right? Unfortunately, it is not that easy. Most of the things you encounter in daily life are very easy to see —large, close, bright, and colorful. Conversely, most of what you'll be seeing through your telescope is small, faraway, dim, and with little or no color. Every stargazer has to get used to the idea of seeing in a different way when looking through a telescope.

One of the difficulties of this different way of seeing is that many factors seem to conspire against getting a good view of the night sky. First, Earth's atmosphere works against the observer. The turbulence of the upper atmosphere makes images in a telescope—already fragile—twinkle and boil on many seemingly clear nights, in much the same way heat rising off a tar road makes the horizon seem to wave. Second, dew can form on the telescope optics on even warmer nights, decreasing the clarity of observations and increasing observer frustration levels. Third, in some areas of the country, completely clear, good seeing nights are few and far between, so it may seem that Nature is deliberately hiding itself from telescopic observation.

However, if these obstacles to good seeing can be avoided, a few other pointers will improve your seeing greatly. One technique to improve your observations is the use of averted vision. Surprisingly, the most sensitive part of the human eye is not the part that you view an object with directly. You'll see more detail if you avert your vision slightly, so that your direct line of vision is

Even in this field of view—a small portion of the sky—there are thousands of stars to see. One of the advantages of having a catadioptric telescope is that the placement of the eyepiece enables the telescope to be mounted high or low, depending upon the height of the observer.

not on the object which is to be viewed. Also, your eyes need to become dark-adapted before they will really be capable of seeing the night sky well. Your eyes in daylight and under artificial lighting are constantly bombarded with many light waves. It takes several minutes for the slight (temporary) burning on your retinas caused by this exposure to strong sources of light (such as an average fluorescent light) to cool. Many stargazers spend fifteen minutes to a half hour in a completely darkened room before observing, letting their eyes get accustomed to seeing in very little light. Some astronomers go so far as to go to bed very early on evenings that they are going to observe and then wake up in darkness. This way, their eyes are already dark-adapted and well rested, and they can proceed immediately to the eyepiece.

Well-rested eyes are perhaps the most important tool for good seeing. If you are relaxed while viewing, you will see many times more detail than even an advanced astronomer who is suffering from eyestrain. There are several ways to avoid eye problems. Averted vision aids in relaxing the eyes. Another trick stargazers use to cut down eyestrain is learning to keep open the eye with which they are not looking through the telescope. If only the viewing eye is kept open, all of the strain of seeing is thrust upon that eye, and eye tension soon results. Another way of avoiding strain is by not overtaxing your eyes. Rest them every few minutes. A good rule of thumb is that every five minutes the eyepiece should be thrown out of focus, the eyes rested for several seconds, and the eyepiece then refocused before any additional viewing is done. One final bit of advice is the old adage, practice makes perfect.

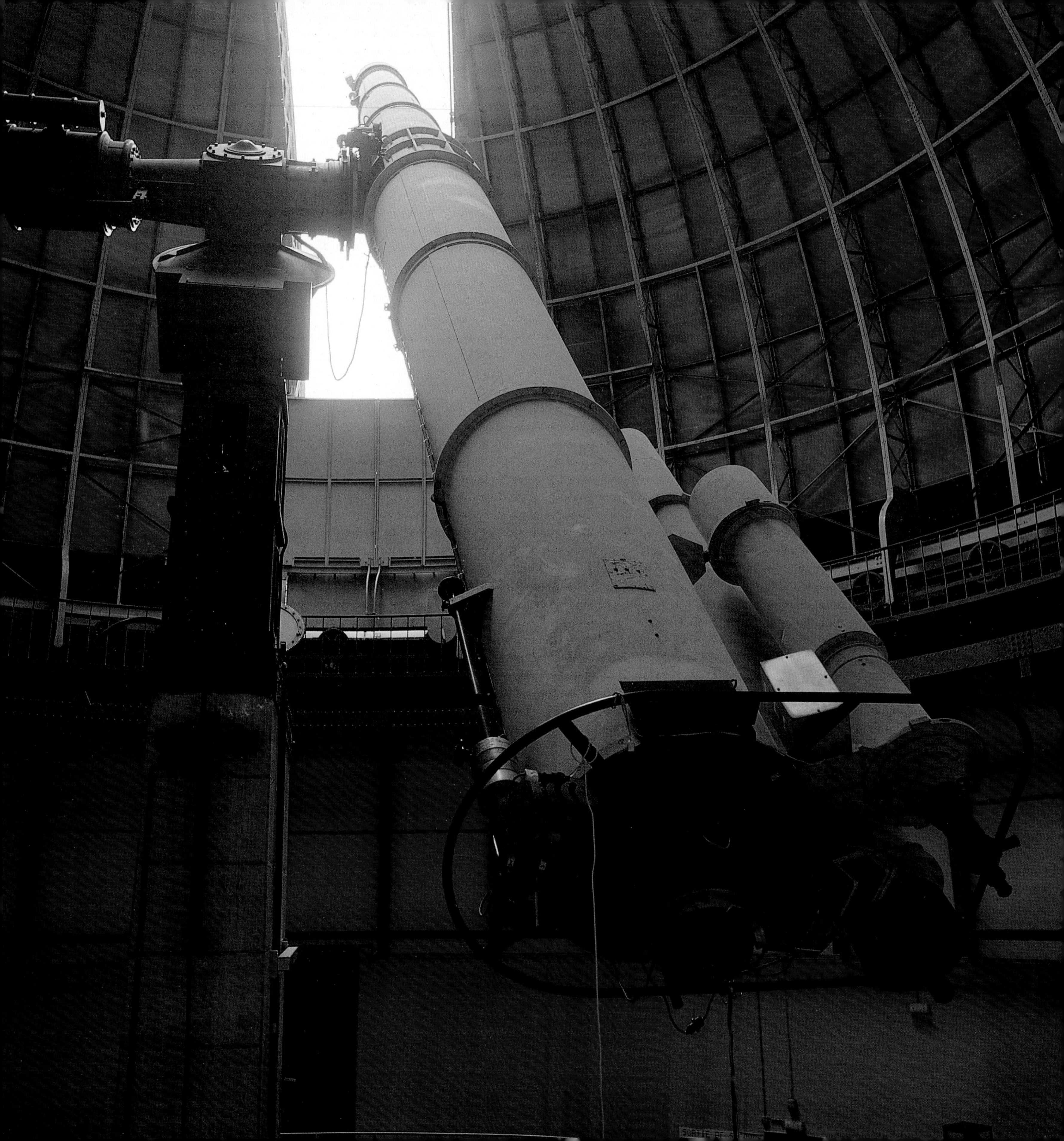

Not many amateur stargazers get a chance to use the huge telescopes housed in observatories like the one pictured AT LEFT. AT RIGHT, in this photograph at dusk, both the moon and Jupiter are in view. Most of the planets are best viewed at dusk and just before dawn. In addition to the visible wavelengths of radiation (light) that allow us to view the stars, many celestial bodies emit other forms of radiation. These radio telescopes pick up radio frequency waves, while other instruments measure ultraviolet, x- and gamma rays from the heavens.

The Giant Scopes

For serious astronomers, the great quest is for more aperture—bigger and more optically perfect objectives and mirrors with which to view ever more distant celestial objects. An amateur astronomer probably won't encounter anything over three feet (one meter) (because of the prohibitive cost and size of large mirrors or object glasses, it's rare to come across a telescope any larger than twenty inches (fifty centimeters) or so). However, some professional astronomers affiliated with government organizations, universities, or major observatories gain access to federal funding to subsidize the tremendous cost of manufacturing ever larger optical telescopes. Here is a list of the largest in the world:

REFLECTORS

Aperture inches/ centimeters	Telescope	Date Opened
236/590	Mount Semirodriki, USSR	1976
200/500	Hale Reflector, Mt. Palomar, USA	1948
174*/435	Multiple-Mirror Telescope, Mt. Hopkins, USA	1979
158/395	Cerro Tololo, Chile	1970
158/395	Kitt Peak, USA	1970

It's said that the cost of building a large telescope is directly proportional to the number of governmental authorities who have to approve the project. It wasn't always so. The first big telescope was a forty-nine-inch (122-cm) reflector built by none other than William Herschel in 1789. It reigned as the world's largest until 1845, when a seventy-two-inch (180-cm) reflector was unveiled by Lord Rosse in England. This, in turn, was biggest until the much heralded Hooker reflector was built on Mt. Wilson in California in 1917. Now, of course, the huge Soviet telescope and the Hubble Space Telescope lead the way in light-gathering capability, but who can say what giant telescope may be built in the future by someone with a dream of seeing ever further.

REFRACTORS

Aperture (inches)	Telescope	Date Opened
40/100	Yerkes Observatory, USA	1897
36/90	Lick Observatory, USA	1888
32.7/82	Meudon, France	1893
32/80	Potsdam, East Germany	1899
30/75	Allegheny, USA	1914
30/75	Nice, France	1880

*The Multiple Mirror Telescope (MMT), as its name implies, uses six mirrors in conjunction to equal the light grasp of a single 174-inch mirror.

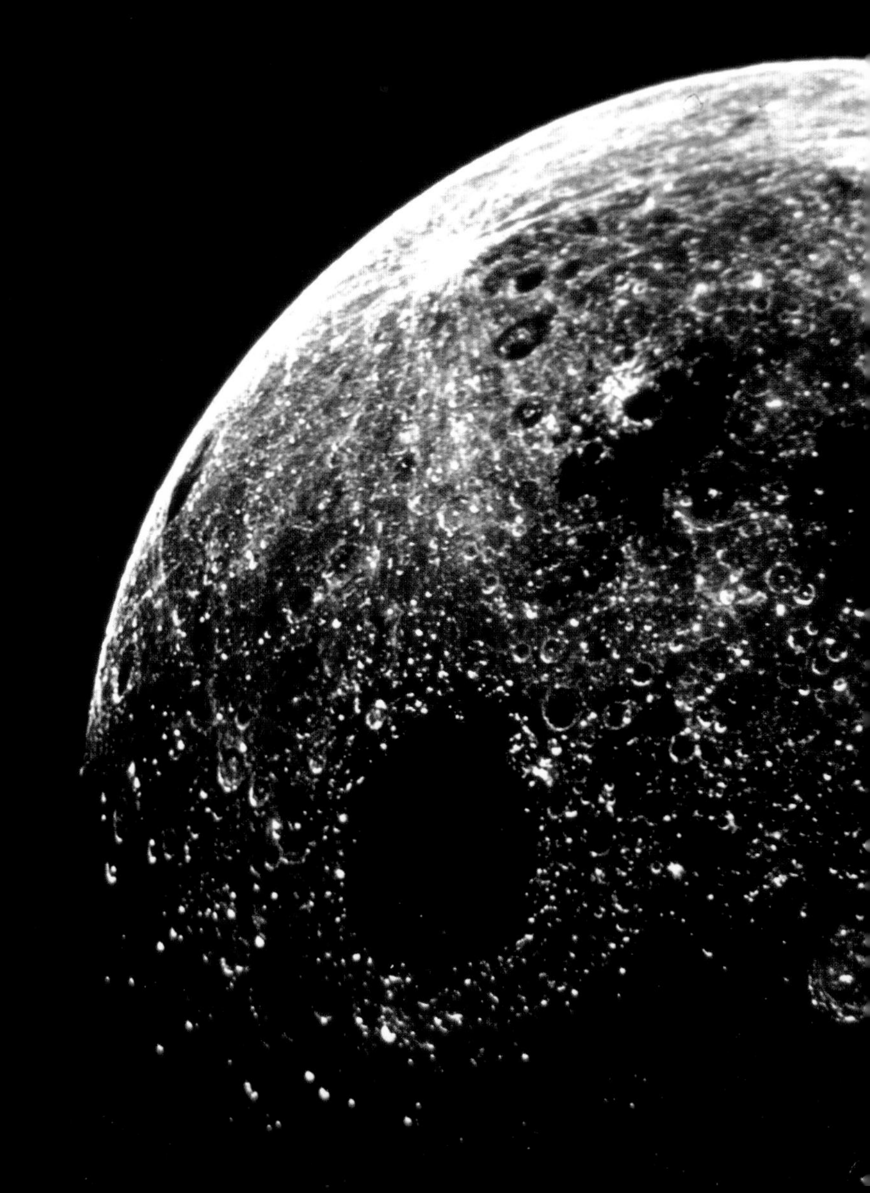

C H A P T E R   S I X

LOOKING AT THE MOON

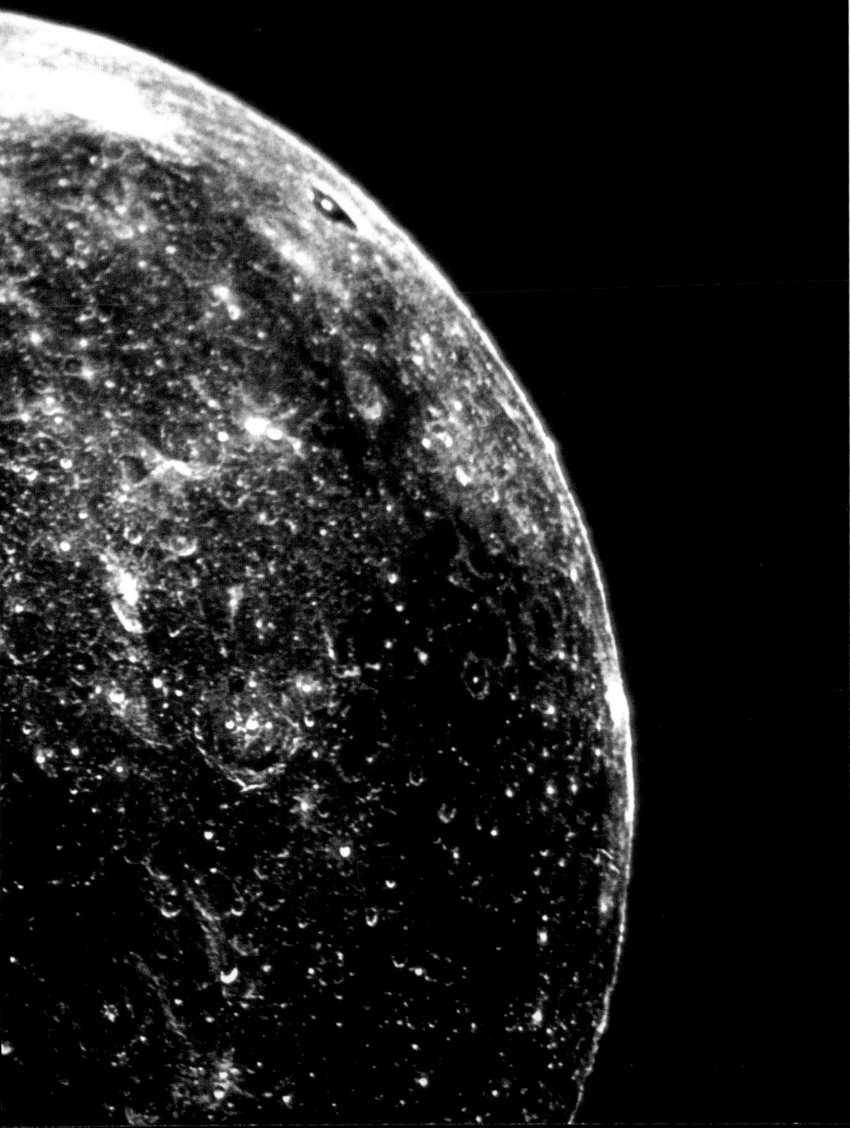

The moon, pockmarked with craters, and the watery Earth. In this spectacular photo, the relative sizes of the Earth and moon are reversed; the Earth is actually six times larger than the moon.

© Brian Parker/Tom Stack & Associates

The moon is the nearest major celestial object to Earth, and perhaps the one that is most responsible for luring new stargazers to explore the heavens. The moon has always been an object of great study and observation. Even the earliest astronomers and cultures placed special emphasis on the place of the moon in the scheme of life.

Modern stargazers are usually attracted by the brightness of the moon, which makes viewing it without optical aids very easy. In fact, many lunar details can be observed by the trained unaided eye.

The aspiring stargazer's first look at the moon with any sort of magnification is very exciting. In a pair of 7 × 50 binoculars, the moon becomes dazzlingly bright and a whole new world of seeing opens up. After years of seeing spectacular photographs of the surface of the moon, the beginning stargazer can get the same view with a relatively inexpensive pair of binoculars. In fact, perhaps the best views of the moon are through binoculars or very small telescopes. Most large telescopes take in too much of the moon's light and don't provide enough of a field of view, although they do allow for very detailed observations of small areas of the moon's surface.

Though you may not believe it when you first view the moon through a telescope, the moon does not emit any light itself—all of its light is a reflection of the sun's light. Therefore, when the sun, moon, and Earth are at angles, the moon appears only partially. The line that separates the illuminated portion of the moon from the dark portion is called the terminator. The majority of moon observations are made on the lighted side, also called the day side, near the terminator. It is in this region that the most detail is apparent. This is because the area near the terminator is less illuminated than the area directly facing the sun, so the features of the lunar surface are highlighted by the high contrast of deep shadows. In fact, many lunar observers find that the moon is excessively bright when it is nearing full and either compensate for this brightness by using a filter on their telescope while observing or don't observe during these periods.

FEATURES OF THE LUNAR LANDSCAPE

The moon has several notable features. Its surface is made up of highlands and lowlands. The highlands are the lighter-colored mountainous areas. The lowlands are called the *maria* (Latin for *seas;* singular is *mare*), because moonwatchers originally thought these large, darker areas were bodies of water. They are easy to see with binoculars and even with the naked eye. The *maria,* and the moon itself, actually contain only very minute traces of water. The *maria* resemble dry desert plains and were formed from lava late in the development of the moon. These lava plains are evidence that the surface of the moon has not changed appreciably in hundreds or even thousands of years.

The highlands are marked by craters, sometimes severely. Almost all of the craters on the moon are now believed to have been formed by the impacts of meteorites, including some that must have been very large. There are also craters in the *maria,* but not nearly so many as in the highlands. This would seem to indicate that the *maria* were formed after the majority of meteorite strikes formed the craters.

Other features of the lunar surface include domes, mountain ranges, and valleys and clefts. Domes appear as small welts on the

ABOVE LEFT, the full moon, haunting and beautiful even on a night of relatively poor seeing conditions. ABOVE, the moon during a lunar eclipse. The shadow of Earth is visible crossing the face of the moon, and the discoloration near the edge of the shadow is due to the reflection of the Earth's atmosphere.

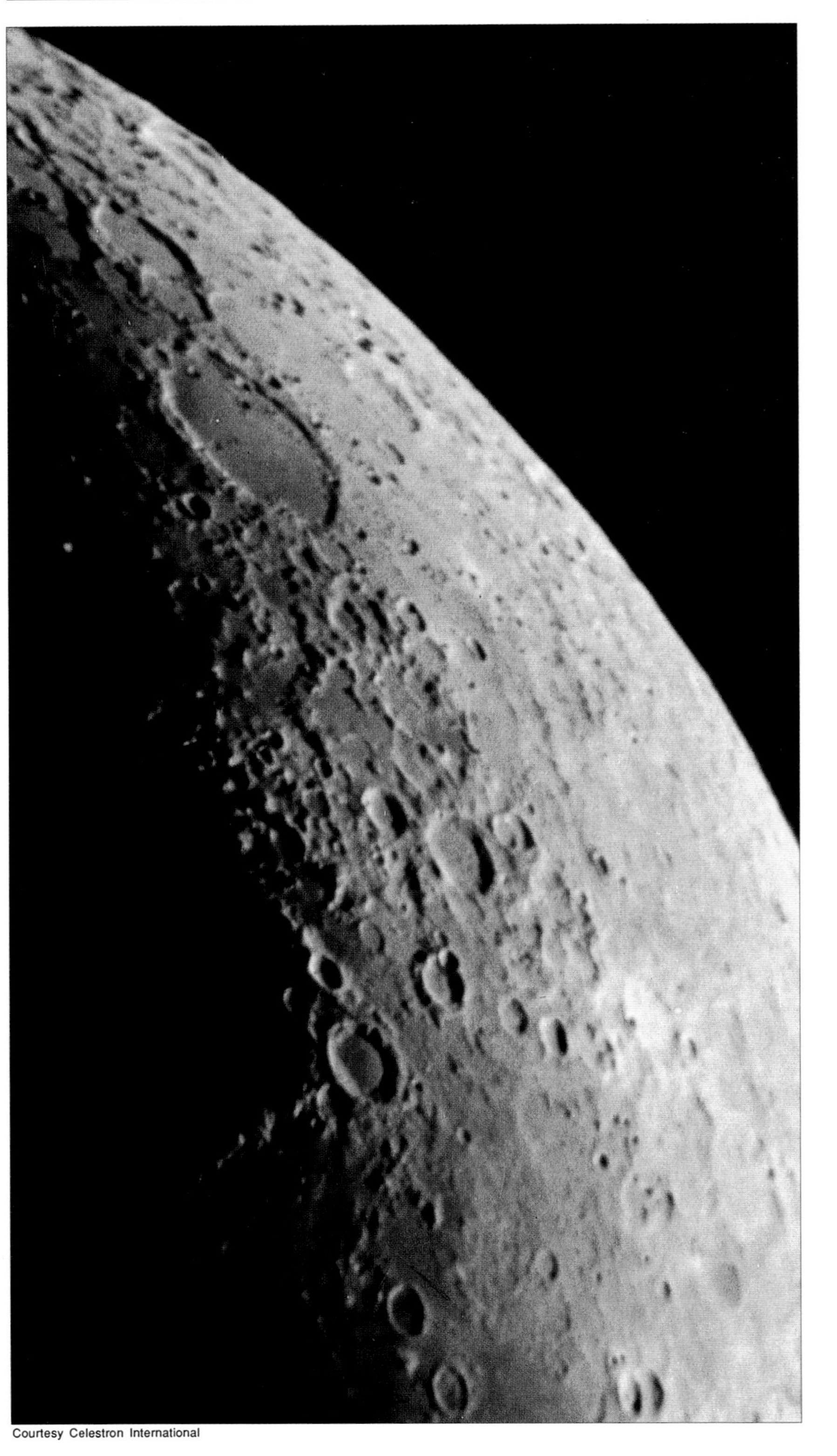

Courtesy Celestron International

© George East

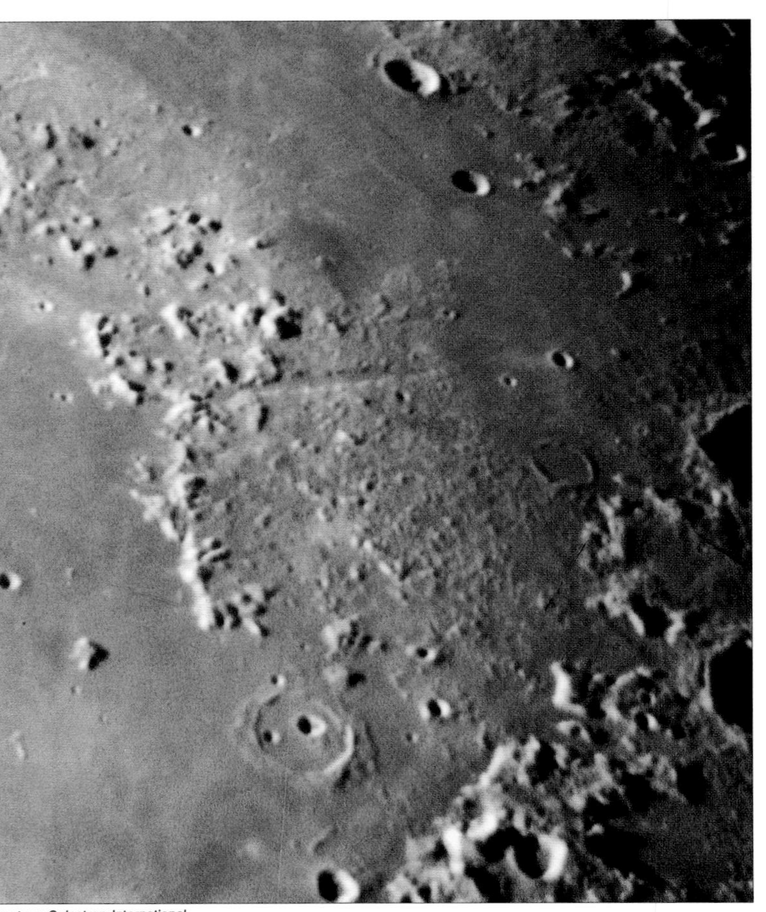

Courtesy Celestron International

Courtesy Celestron International

Several views of the lunar landscape show the variety of features on the surface of the moon. Great plains surrounded by mountains and rilles remain relatively free of craters, while other areas are densely packed with deep and wide craters. Scientists still cannot explain the genesis of some of these features.

face of the moon, smooth and round rather than craggy and irregular. Their genesis is still not completely understood. Mountain ranges cut across the moon in several places and are the result of both volcanic activity and shifting of the moon's crust. The highest mountain peaks reach over thirty thousand feet (nine thousand meters), higher than Mt. Everest (on a satellite that is many times smaller than Earth). Lunar valleys and clefts, also called rilles, cross the surface of the moon and range from a few hundred yards long to huge gashes several hundred miles long and several miles wide. They may be due to craterlike impressions, or they may be caused by shifting cracks just below the surface of the moon.

In addition to these features, within the past thirty years there have been reports of "events" on the moon. Though there is no photographic confirmation of events of any kind, it is reasonable to believe that at least some of the reported sightings are accurate. Called transient lunar phenomena (TLP), the reported events range from a brightening or pulsating inside a crater to flashes of bright or colored light on the lunar surface, usually in the *maria*. Many of these TLPs may be caused by observer eyestrain rather than an event on the moon, but there are many experienced amateur and professional astronomers who swear to their observations of TLPs. As of this writing, there is no positive explanation for TLPs, though the leading theory is that they are caused by discharges of various gases from beneath the moon's surface. The emissions may either cause the TLPs by themselves or by the way light is reflected off of the clouds of gas as it disperses. Though it doesn't pay to look specifically for TLPs—invariably the eyes see an event because they want to see an event—it is valuable for the lunar observer to keep an open mind.

Courtesy Celestron International

© George East

AT LEFT, the edge of the visible portion of the moon when it is not full is called the terminator. The best observations of the lunar surface are made near the terminator, where there is less glare. ABOVE, the beauty of the moon at twilight is unparalleled. Dusk is an excellent time to view the moon, even on nights when the seeing isn't good. In fact, many of the most useful observations of the moon have been made with the naked eye on hazier nights, when glare doesn't strain the observer's eyes.

LUNAR OCCULTATIONS

Another lunar phenomenon that is interesting to observe is an occultation. An occultation, or hiding, occurs when the moon passes in front of a star or planet, obscuring it from view. If the star or planet passes completely behind the moon, it can be timed from its disappearance, called immersion, until its reappearance, called emersion. If the time is calculated with the exact known position of the observer on Earth, the precise position of the star in the celestial sphere can be determined. Proof that the moon has no atmosphere can be observed during a lunar occultation: There is no flickering or wavering of the star as it nears the edge of the moon. The occulted star will suddenly and very clearly disappear from view when it encounters the limb of the moon, particularly if the star disappears behind the darkened limb before a full moon.

Another type of lunar occultation, a grazing occultation, occurs when a star just grazes the edge of the moon in passing it. As the star passes the moon, it may disappear and reappear several times. This is not because of any lunar atmosphere, but rather because the star passes behind mountain peaks and becomes visible as it crosses valleys on the edge of the visible moon. If these observations are recorded accurately, a map of the particular section of the lunar surface can be drawn very accurately, including heights of peaks and distances of lunar features.

If occultations interest you, an organization called the International Occultation Timing Association (IOTA) can alert you to upcoming occultations and help you get started making serious observations of lunar (and other) occultations. They require a small membership fee and request accurate and precise data from observers. Becoming an active member of IOTA is one of the ways an amateur astronomer can make contributions to the astronomical pool of knowledge. Information on upcoming occultations is also readily available in *Astronomy* and *Sky & Telescope* magazines.

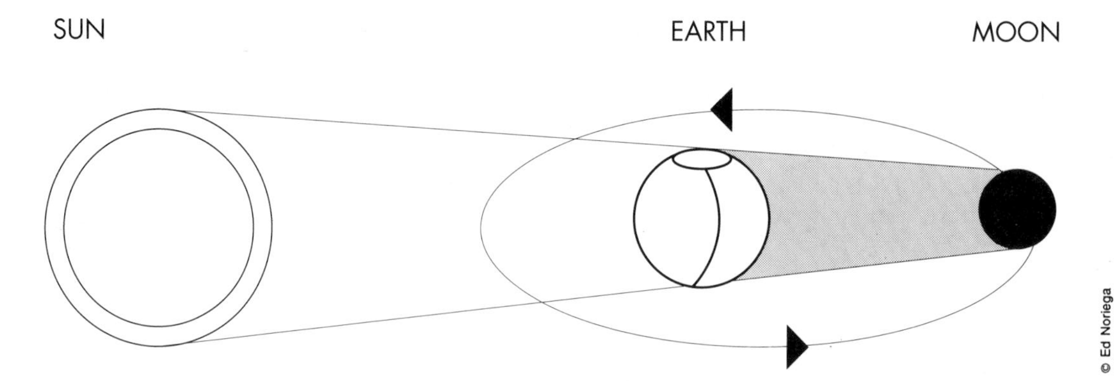

© Ed Noriega

Lunar eclipses are among the most fascinating celestial events. The diagram shows how the sun's light is blocked by the Earth, causing an eclipse. BELOW, the moon nears full eclipse. AT RIGHT, the stages of a partial eclipse are captured over a span of about four-and-a-half hours.

LUNAR ECLIPSES

Perhaps the most spectacular lunar event, a lunar eclipse, is really an Earth event. A lunar eclipse occurs when the moon passes through the shadow cast by Earth. There are two types of lunar eclipses: total eclipses and partial eclipses.

When the entire moon falls into the earth's shadow, a total eclipse occurs. The previously full moon is darkened for up to one hundred minutes as it sweeps past the Earth. In a perfect solar system, the moon would completely disappear from view during a total eclipse; however, because the rays of the sun are bent by Earth's atmosphere, some light continues to fall on the moon during a total eclipse, and it takes on a faint dark-red glow. When a total lunar eclipse occurs, it is visible to anyone for whom the full moon was visible in the sky that evening (contrast this with solar eclipses, which are visible only to a very small area in varying degrees depending upon the location of the observer on Earth at the time of the eclipse).

A partial lunar eclipse occurs when a part of the moon passes through the Earth's shadow. In this type of eclipse, only a portion of the moon's surface is darkened, and then only for a short time. These eclipses are interesting for the observer because areas of the moon may change color, and there may be unusual views, particularly if you are looking through binoculars or a small telescope.

Courtesy Celestron International

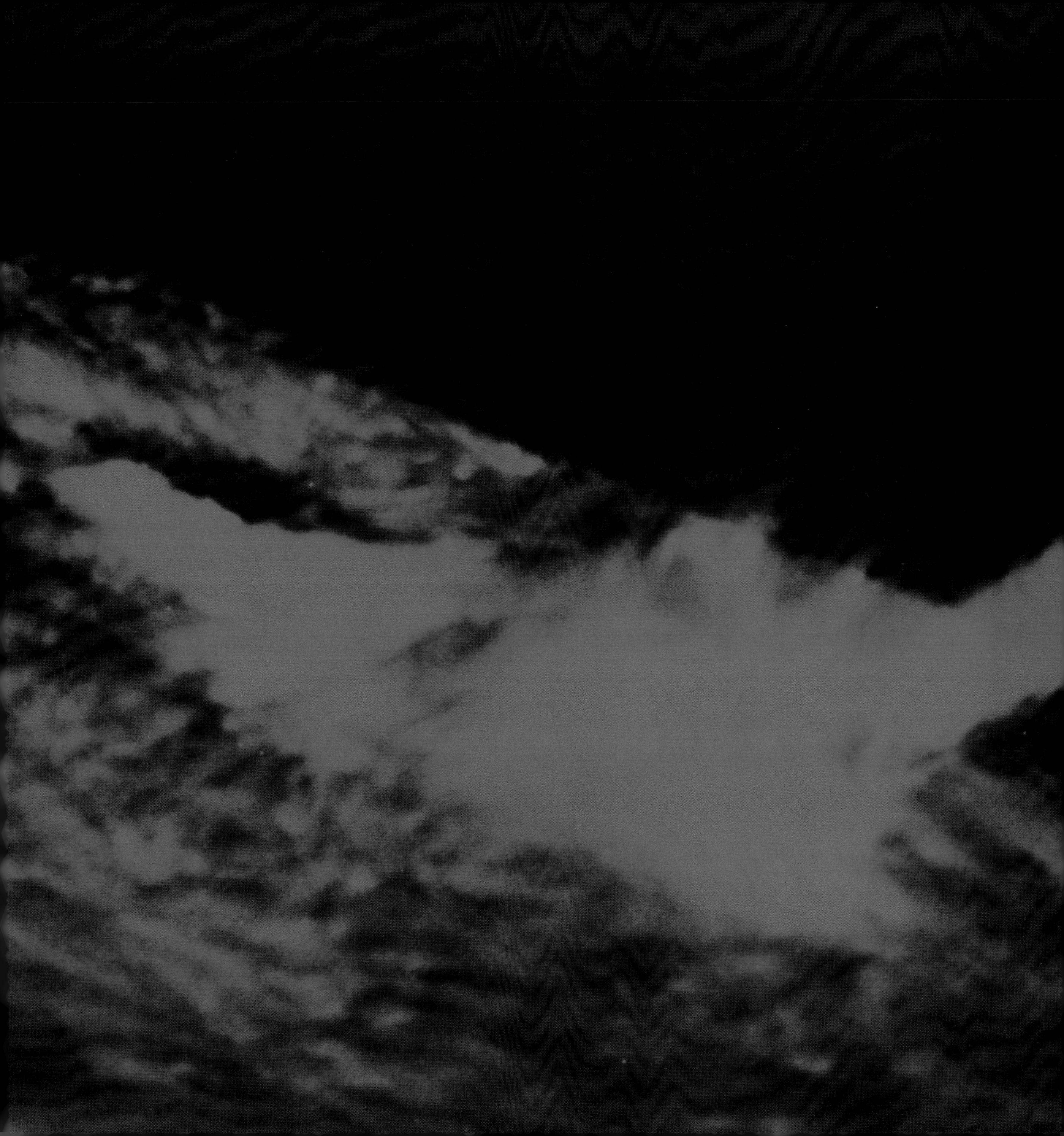

C H A P T E R   S E V E N

LOOKING AT THE SUN

Courtesy National Optical Astronomy Observatories

A solar flare on the surface of the sun. These spectacular sights should only be viewed indirectly, or under the supervision of trained solar astronomers.

ABOVE, this is the right way to view a solar eclipse. Here, a scientist is showing sunspots to a group of onlookers in Rangapur, India. AT RIGHT, *the sun is often viewed at sunrise, when its rays are less dangerous. This is because they must travel through a much thicker layer of the Earth's atmosphere, and are somewhat dispersed.*

Any discussion of viewing the sun must begin with a warning: *Never look directly at the sun with an optical instrument!* The sun's rays, magnified through a pair of binoculars or a telescope, can cause blindness in just a fraction of a second. Many manufacturers offer solar filters for use in viewing the sun through a telescope, and these filters typically allow less than one one-hundredth of one percent of the sun's light to penetrate the filter. Even this may not be adequate to prevent permanent scarring of the retina, and the great heat generated by the amount of light gathered in a telescope's objective can damage the optics of the telescope. If you must view the sun, be sure to look at a projected image, and be very careful when using your telescope.

The way to project the image of the sun is to mount a piece of white paper behind the eyepiece of your telescope or binoculars. To project the entire disk of the sun, use very low power. The distance of the paper from the eyepiece can be adjusted to give the correct image size. Only on this white sheet can the features of the sun be safely viewed. In fact, one advantage to this method of observation is that you can easily sketch the apparent features on the paper itself, thereby preserving the image for later comparison with other images. Shield the projection paper from direct sunlight outside the telescope; the only light hitting the paper should come from the eyepiece. You can do this by attaching a large piece of cardboard or other screen around the outside of the telescope, so that the projection paper lies in shadow except for the image. Aim the telescope directly at the sun by watching the floor behind it: When the shadow cast by the telescope itself is smallest, then it should be aligned directly with the sun. This position should coincide with a noticeable burst of sunlight emerging from the eyepiece.

The most common feature of the sun is sunspots, dark spots on the "surface" of the sun (this apparent surface is called the photosphere, but the sun is actually a ball of gases and thus has no surface). In addition to sunspots, other features may show themselves on the projection paper. Faculae, bright clouds above the photosphere, signal the presence of sunspots; they appear just before a sunspot forms on the photosphere. Granulation, the

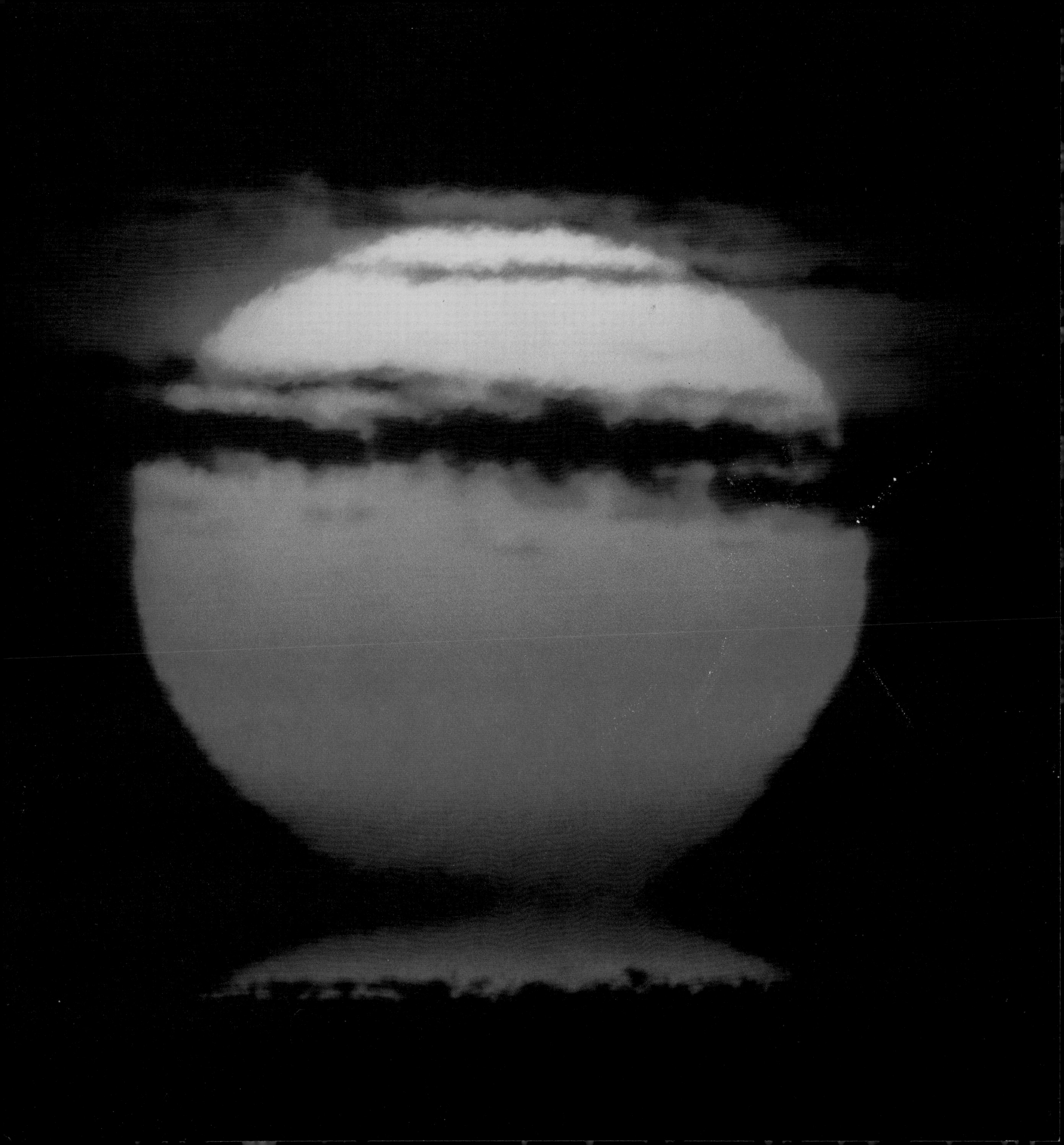

Courtesy NASA

"texture" of the photosphere, is usually discernible only in telescopes with apertures of 100 mm or greater and very high magnifications. Solar flares are eruptions on the face of the sun lasting only minutes. They are among the sun's most stunning and rare displays.

Solar observations are particularly interesting if they are made over several days or weeks and if they are accurately recorded in drawings. Plotting the course of sunspots across the rotating photosphere and then awaiting their reappearance on the opposite side can be very rewarding and instructive, and will leave an observer with a visual testament of the activity of the heavens.

Both ALPO and AAVSO have solar groups that ask for observations and offer instruction in making specific examinations of the sun. Addresses for both can be found in the back of this book.

SOLAR ECLIPSES

A solar eclipse is by far the most dramatic event that the heavens provide. During a solar eclipse, the sky darkens quickly and the familiar light of the sun is snatched away by the moon, leaving a dark hole in the sky. Throughout history, solar eclipses have been heralded as portents of great change. Solar eclipses have even been credited with stopping two ancient wars whose participants felt the eclipse was a sign from the gods to stop fighting. Nowadays we know that solar eclipses are regular, predictable events and not messages from the gods, but they are awesome spectacles nonetheless.

Solar eclipses are caused by the moon passing in front of the sun, blocking its rays from reaching Earth. This is possible because the sun and moon cover about the same amount of our

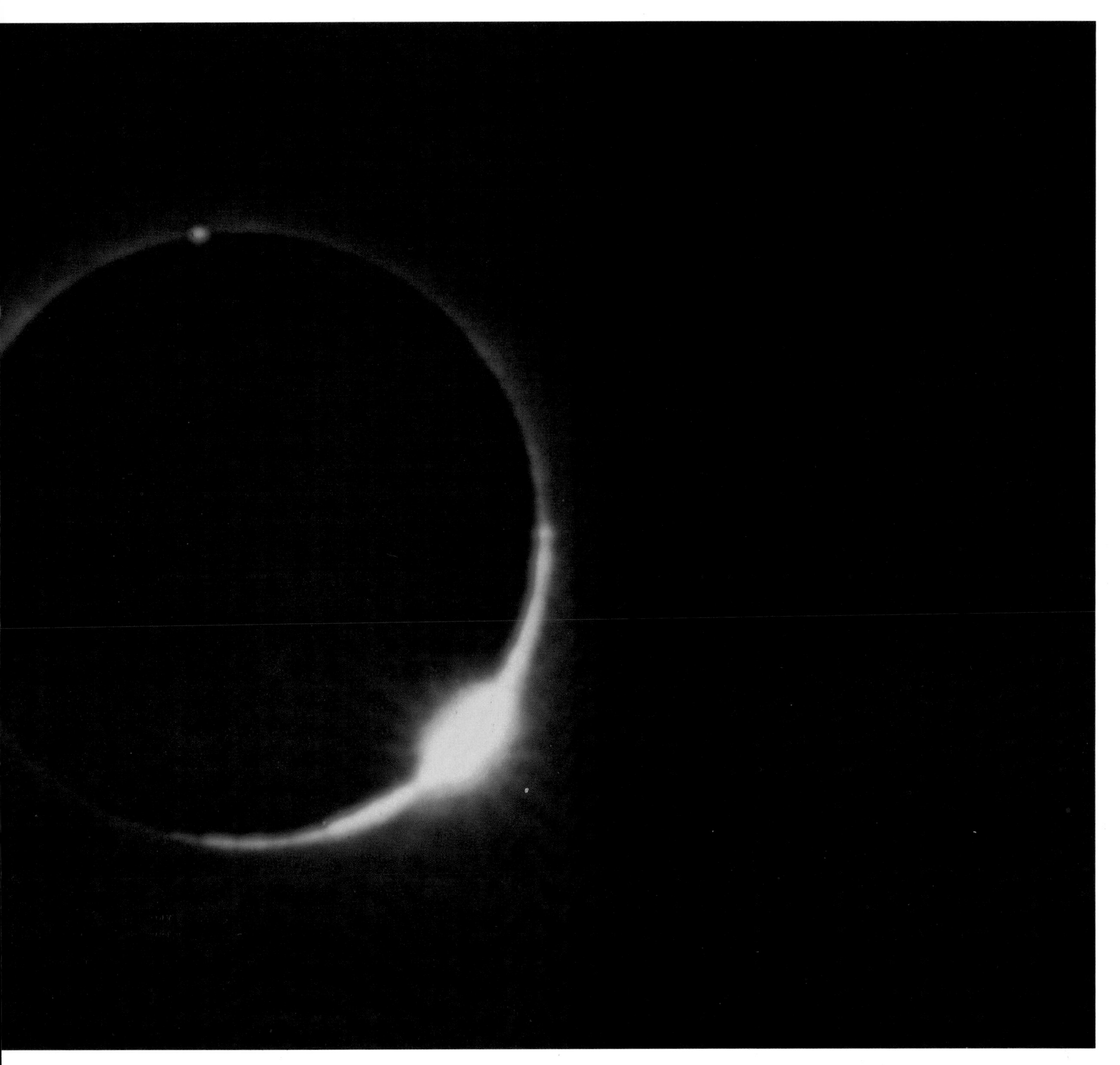

The sun's corona is visible just after a total solar eclipse. During the short time that the view of the sun itself is blocked, astronomers can study the makeup of the atmosphere around the sun, and often view eruptions of the surface.

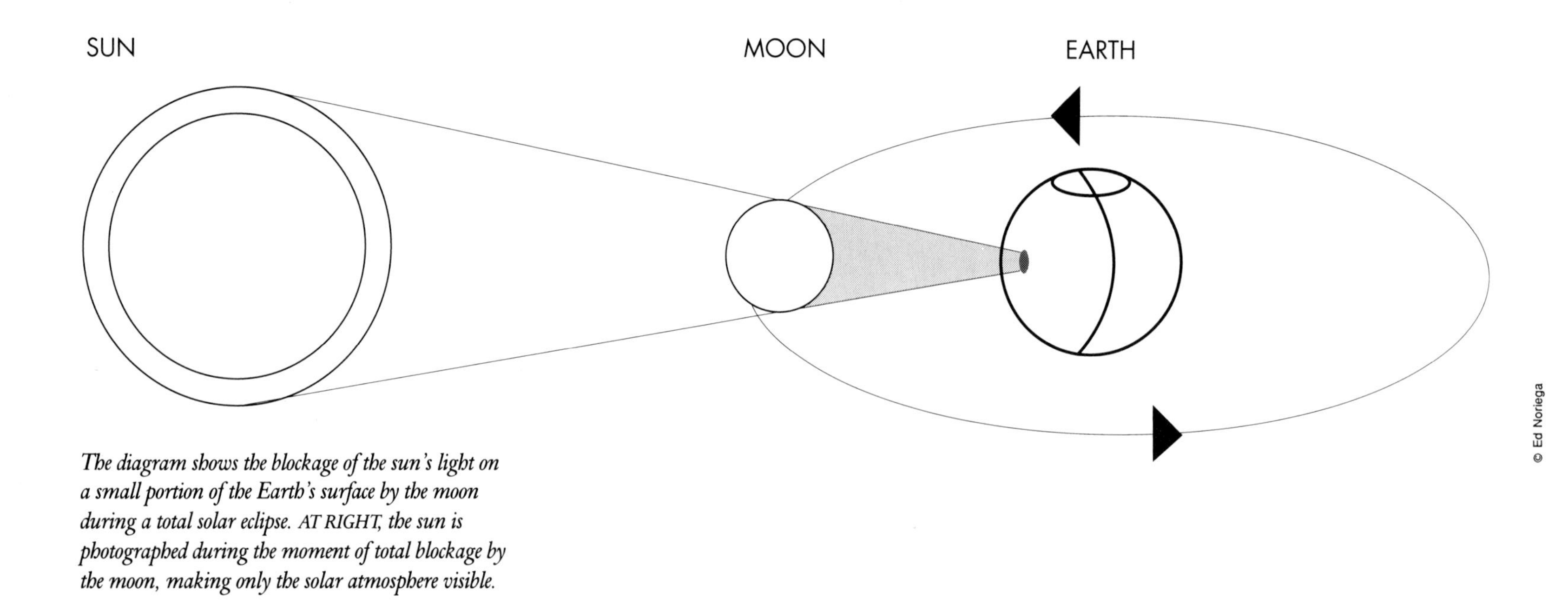

The diagram shows the blockage of the sun's light on a small portion of the Earth's surface by the moon during a total solar eclipse. AT RIGHT, the sun is photographed during the moment of total blockage by the moon, making only the solar atmosphere visible.

sky. From our view on Earth, they are the same relative size.

During a solar eclipse, the shadow cast by the moon passes across the Earth. For most areas of the Earth, the eclipse is not apparent at all. In many other areas, the eclipse is only partial—the moon blocks only a portion of the sun. However, areas in the umbra of the eclipse receive the entire shadow of the moon. They are lucky enough to be in correct alignment for the eclipse, and for them, the eclipse is total. In this case, the sun's corona—its atmosphere—is visible around the darkened disk of the moon. This total eclipse is visible for only a few minutes. Total eclipses only happen about once a year, and since there is usually a very limited viewing area for a solar eclipse, they are rare sights for most people.

Partial eclipses are more common for two reasons. First, whenever there is a total eclipse, only those in perfect alignment see it as a total eclipse. Those who are just out of the totality, or full darkening of the eclipse, see a partial eclipse, and the partial eclipse is visible over a greater area than the totality. Second, sometimes the moon is too far away from Earth for its umbral shadow to fall on Earth. We receive the penumbral half-shadow during these true partial eclipses, but there is no totality on Earth.

Solar eclipses are slightly more common than lunar eclipses. However, while the solar eclipse only strikes a limited viewing area, a lunar eclipse is visible to all the people who are in the dark hemisphere of Earth during the event, making it a much less rare phenomenon.

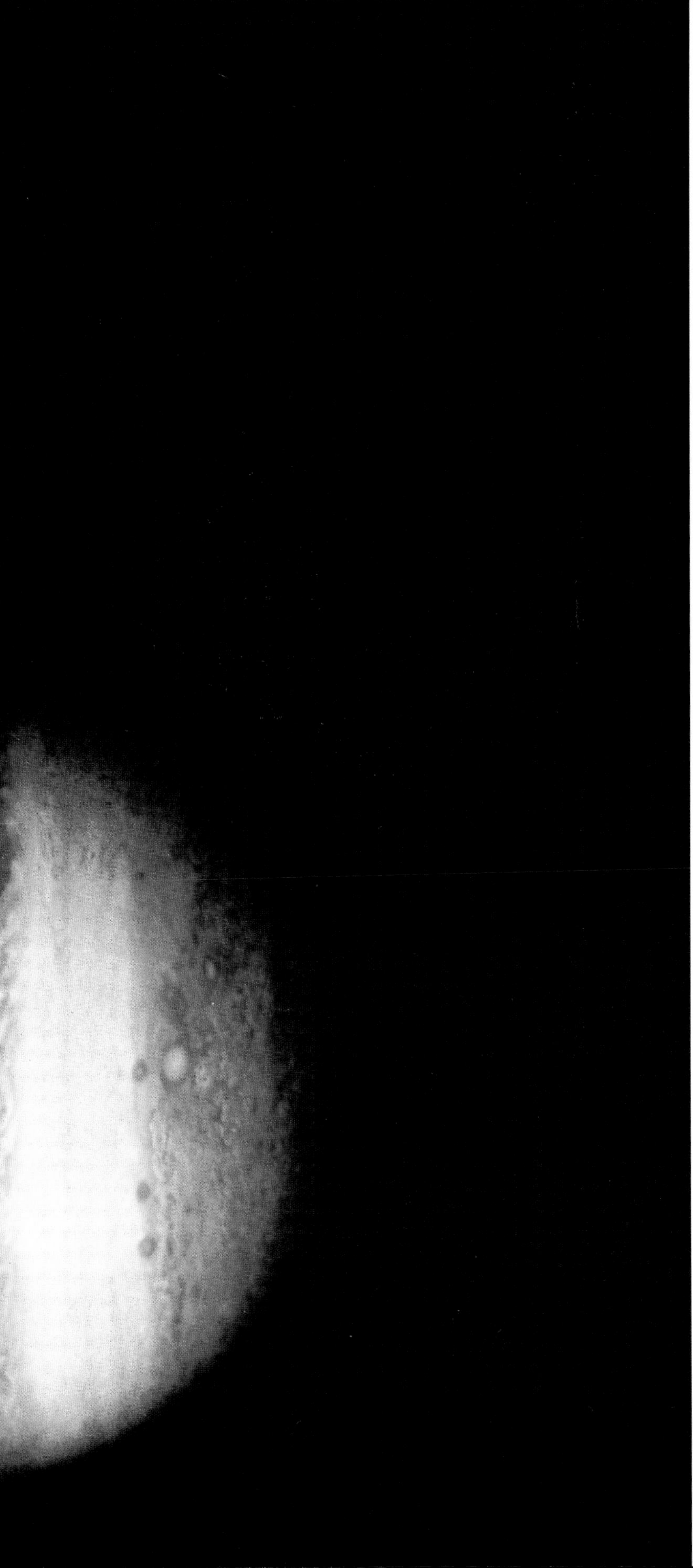

C H A P T E R   E I G H T

VIEWING THE PLANETS

Courtesy NASA

Four of Jupiter's moons are assembled in this montage. While not to scale, they are in their relative positions from Jupiter. From nearest to farthest: Io, Europa, Callisto, and Ganymede.

An artist's conception of Pluto and Charon. Scientists now hypothesize that Pluto and Charon may be a binary system, rather than a planet and its moon.

Except for the sun and moon, the planets are the celestial objects that appear largest to observers on Earth. Though they look like stars when viewed with only the naked eye, even ancient observers were able to differentiate between the so-called fixed stars and the planets, five of which are visible without optical aids. During the centuries when Ptolemy's cosmology prevailed, it was believed that the five known planets formed the five outer rings of the cosmos, between Earth and the sphere of fixed stars.

We know now, of course, that there are nine major planets, and that they revolve around the sun, not Earth. Amazingly, Pluto, the most distant planet from Earth, was only discovered in 1930, though its existence had been theorized before then. The five planets known to the ancients—those that can be observed even without binoculars at various times of the year—are Mercury, Venus, Mars, Jupiter, and Saturn. Uranus, Neptune, and Pluto are all fairly recent discoveries, only visible with telescopes.

The major planets can be broken down into two groups. The terrestrial planets are the four nearest the sun—Mercury, Venus, Earth, and Mars. They are all solid orbs, and are all relatively small in relation to the next four planets in the solar system. Those four large planets—Jupiter, Saturn, Uranus, and Neptune—are called gas giants because they are composed mainly of gases and, as such, have fairly low densities and turbulent atmospheres. Until recently the gas giants were rather mysterious to us, but the travels of the space probes *Voyager 1* and *Voyager 2* have vastly increased our knowledge of these planets.

The ninth planet, Pluto, is a recent discovery. We know very little of Pluto, except that it is considerably smaller than the other planets and does not behave like any of them. Except for Pluto, each of the planets of our solar system can be observed from Earth.

ABOVE, the view from the famous Walden Pond on a particularly serendipitous day. Both the crescent moon and the tiny speck of Mercury are visible in the twilight, a rare event indeed. At this time of day, the rest of the moon is faintly visible due to the reflection of light from Earth.

MERCURY

Unfortunately for the planetary astronomer, Mercury is too great a challenge for Earth telescopes during much of the year. However, for very short periods (depending upon your viewing latitude, two to six weeks per year), Mercury is visible to the naked eye. Without optical aids it looks like a moderately bright reddish star, and it dances above the horizon just before daybreak, when it is visible. This is because Mercury is always very near the sun. Binoculars will not significantly improve the view of Mercury, even under optimal conditions. Indeed, the best observations of Mercury are made with larger telescopes, in daylight, when Mercury (like the sun) is quite high in the sky. This type of observing, however, requires some specific planning for the stargazer.

For one thing, great care must be taken when observing in daylight not to let the sun too near the field of view. Mercury in daylight is certainly beautiful, but definitely not worth eye dam-

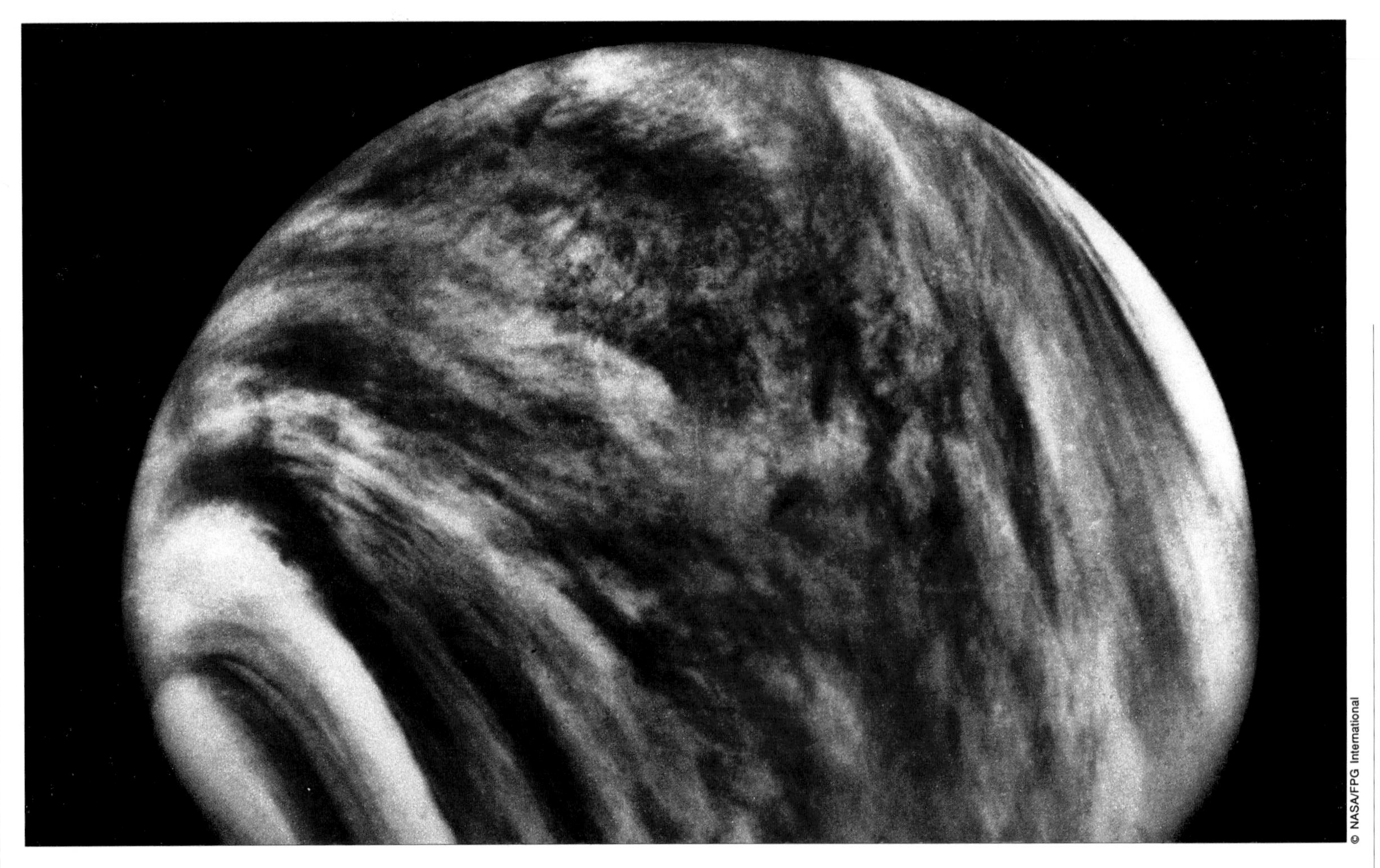

Though Venus is the most conspicuous object in the sky at certain times of the year, its cloud-covered atmosphere prevents us from viewing its surface.

age. A high power is also necessary to enlarge Mercury to a small but visible disk (keep in mind that Mercury is the smallest planet in the solar system except for Pluto). Like the moon, Mercury can be seen in phases, as only the portions of it that are illuminated by the sun are visible from Earth. Finally, because of Mercury's modest size and great distance from Earth, telescopes of at least eight inches of aperture are necessary to give a decent view, and even larger telescopes are recommended.

The surface of Mercury is not unlike that of the moon, barren and scarred. In fact, in size and makeup, the moon and Mercury are surprisingly similar. So you can see why viewing Mercury is difficult—its small size coupled with its great distance from Earth multiplied by its proximity to the sun make it a worthwhile challenge.

VENUS

Venus is known as both the sister planet of Earth and as the shrouded planet. It is close in size to Earth, but that is its only real sisterhood with our planet. Venus is shrouded by thick clouds that fill its atmosphere, and thus the actual planet surface is never visible, even through very large telescopes. However, because it is the nearest planet to Earth, it is easily visible to the naked eye. Venus is often called the morning star because it is visible after sunrise during certain periods of the year. In fact, as with Mercury, excellent observations can be made of Venus in daylight.

To the naked eye, Venus appears as a very bright star, and because it is closer than Earth to the sun, it is easiest to observe

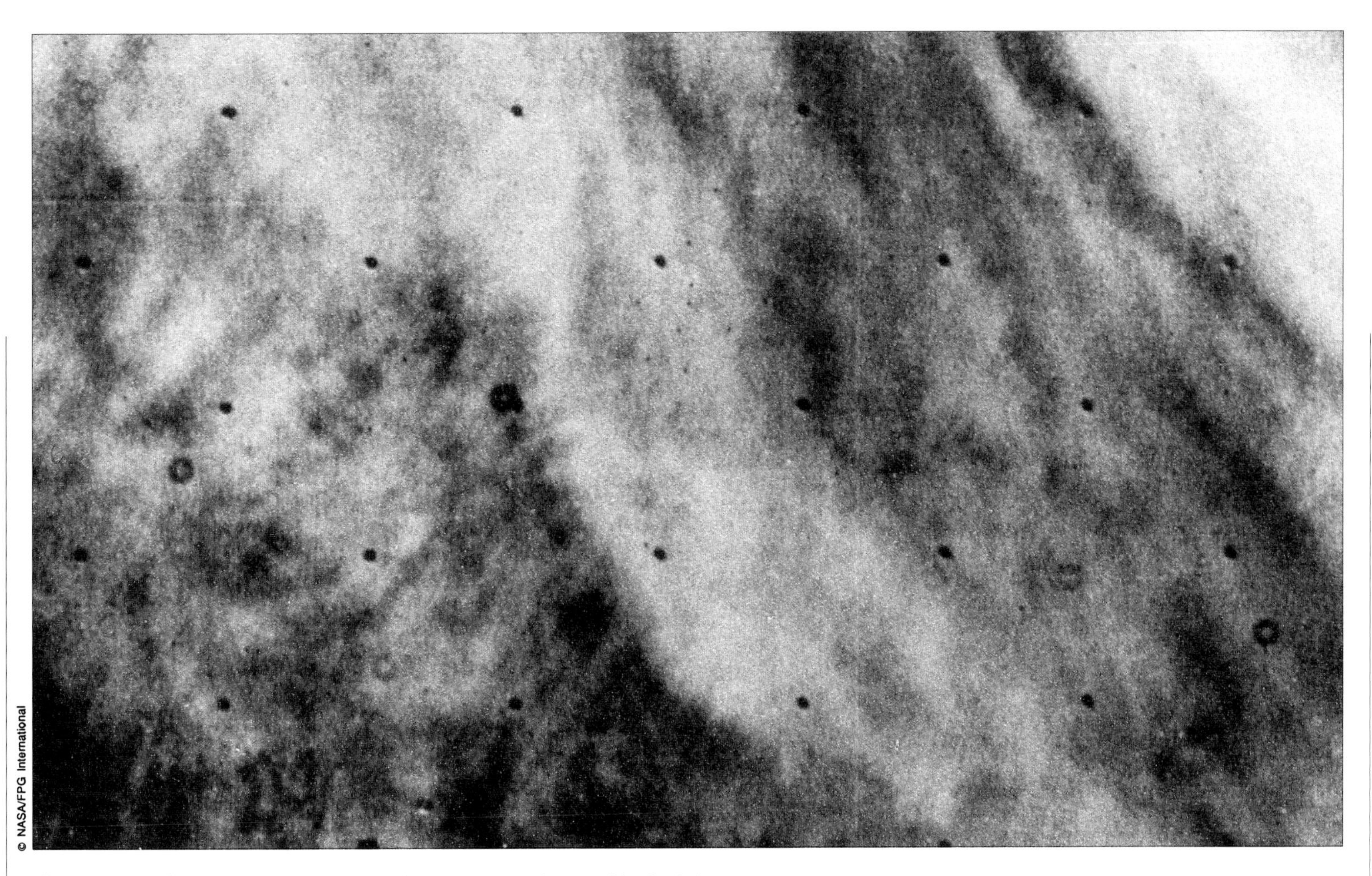

The atmosphere of Venus is, for the most part, carbon dioxide, though some of the clouds that cover it consist of sulfuric and hydrochloric acid. Atmospheric pressure on Venus is nearly ninety times that of Earth.

close to sunrise or sunset, depending upon the season. In a pair of binoculars, Venus is dazzling: It is not quite resolved as a disk, but rather is the brightest point of light in the sky, like a superstar.

In a telescope, the planet takes a shape. Since it is an inferior planet (closer to the sun than Earth), phases are visible at moderate powers. The planet's cloudy atmosphere reflects the sun's light quite strongly, and in fact may be the cause of a peculiar optical trick that has been sighted by astronomers from time to time while gazing at Venus. It seems that some viewers have seen the darkened crescent of Venus glow slightly. Called the Ashen Light, this may be an optical illusion (because it is not universally reported), or the high degree of reflectivity of the planetary atmosphere may create it. Whatever the cause, the Ashen Light is another of those telescopic views that must be carefully observed, for it is easy to see what your mind expects to see, whether it is really there or not.

The best time to observe Venus with a telescope is just before sunset or after sunrise. It is not nearly so visible to the naked eye then, but the view through a telescope will be much clearer and less subject to the optical tricks that can occur when observing a very bright planet in dark skies. Of course, more aperture gives a better view, but Venus can be resolved easily even in three-inch telescopes at moderate to high power, and larger scopes at slightly lower powers.

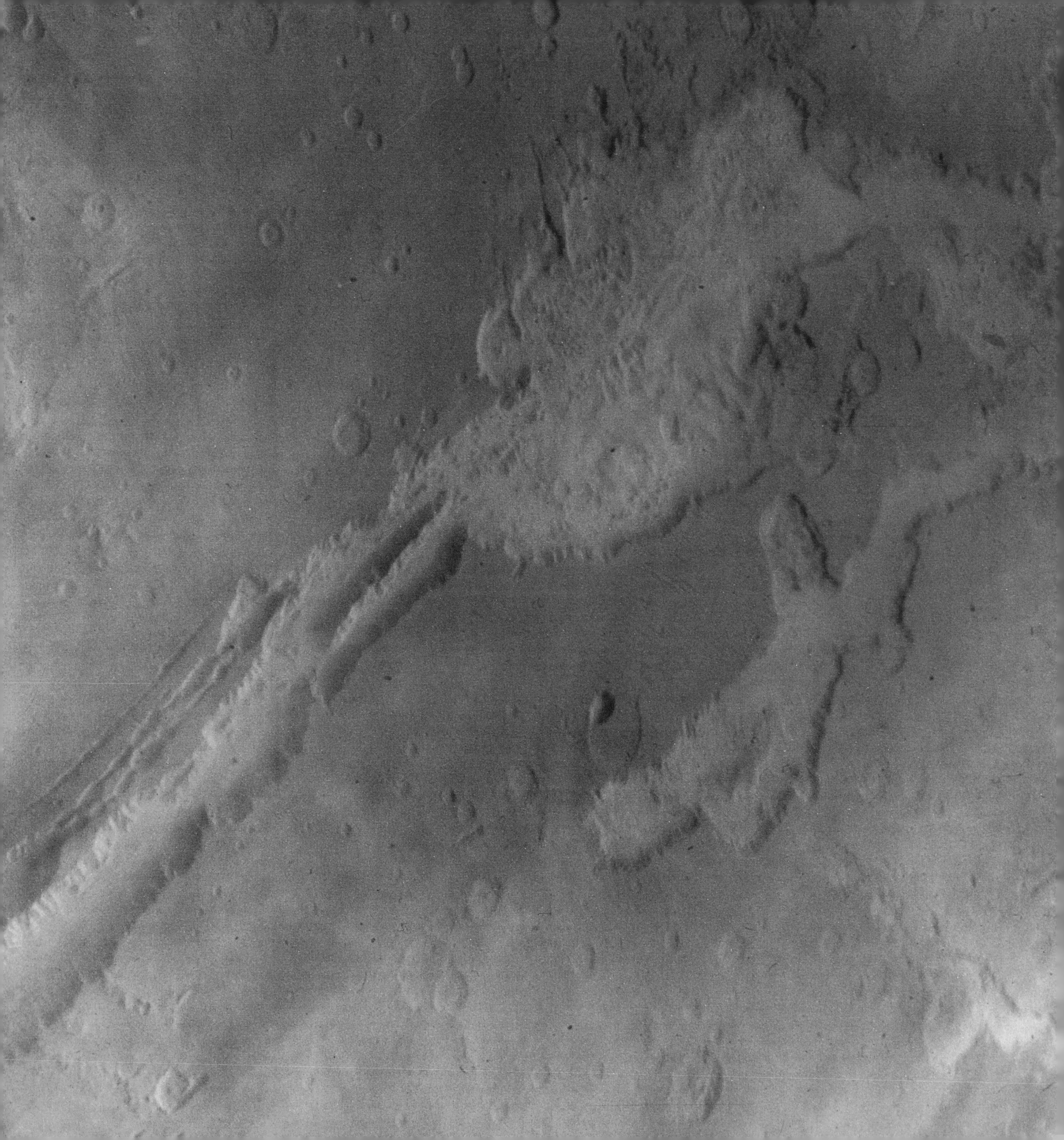

Courtesy NASA

These photos of Mars were taken by Viking 1. AT LEFT, a close-up of the irregular surface of Mars. ABOVE, a composite of three photos taken through color filters and computer-corrected to show the planet as it would appear to an approaching visitor.

MARS

Mars is a remarkable planet littered with meteorite debris and highlighted by huge volcanoes and ravines that may have been formed in its distant past by tremendous rivers of water. It has polar ice caps of frozen carbon dioxide, and appears as an orange-red ball in a telescope. Mars is visible to the naked eye and moves steadily through the zodiac. The view from binoculars is of a reddish star, and even in a small telescope Mars is moderately difficult to resolve into a disk, except at high power. Larger telescopes and higher powers make Mars's markings visible and show off its irregular, ever-changing face.

Through a telescope, several striking features of Mars are clear. One of the polar ice caps is easily visible at all times, and it is quite apparent in larger apertures. The so-called dark areas of Mars appear through a medium-size telescope to be about the same size as the moon's *maria* appear to the naked eye.

The observer benefits from extended observations of Mars. Watching the changing view from night to night is excellent eye training, and Mars's markings, often elusive even to the experienced observer, are much more readily apparent if Mars is a familiar object. Mars has two small moons, Phobos and Deimos. Though they are of about twelfth magnitude and should be easily visible with six-inch telescopes, they actually are difficult objects for telescopes of twice that aperture because the brightness of Mars obscures them.

© Donald Pearson

AT LEFT, the surface of the moon would be an excellent spot from which to observe the other planets and the stars. As you can see, celestial objects are visible to the edge of the horizon, indicating that there is no appreciable atmosphere. ABOVE, this long exposure photograph shows the movement of the stars and a couple of meteors. The meteors are the lines that are not traveling in a circular path. During a meteor shower, often one or more are visible.

THE MINOR PLANETS AND METEORS

In between Mars and Jupiter, in the separation between the terrestrial planets and the gas giants, a belt of small, solid objects orbits the sun. These are asteroids, also (more correctly) called the minor planets. Most are very small, and even the largest can only be resolved by a very large telescope. The brightest asteroid, Vesta, is only of magnitude 5.5 and is thus very dim to the naked eye. The largest asteroid, Ceres, brightens only to the seventh magnitude, and is just 480 miles (768 kilometers) across. It appears considerably dimmer than the smaller Vesta probably because it is composed of less reflective material. Despite their small size and distance from Earth, several asteroids besides Vesta are visible to the naked eye, and nearly fifty are visible in 50-mm binoculars. Several hundred are visible in even the smallest telescopes.

"Asteroid" is a misnomer; the name means "star-like object." The asteroids are actually pieces of matter that in the far distant past may have been destined to be a very small planet. Jupiter's gravity, however, must have sucked up the majority of this matter and dispersed the rest of it into the asteroid belt. The orbits of the minor planets vary, and at certain times of the year, the orbits of some of the asteroids come close to converging with Earth's orbit.

Meteors, often called shooting stars, are asteroids or particles of comets that enter Earth's atmosphere. Usually, they burn up in the upper layer of thin air. Sometimes, however, meteors actually travel all the way to the surface of Earth, burning across the sky and landing as objects smaller than pebbles. In fact, most of the meteorites in the sky are the size of grains of sand and are visible only because they flare up and burst into flames upon entering our atmosphere. On any given night, about six meteorites are visible per hour, though during meteor showers there can be fifty or more. Though meteors are interesting to view through telescopes, the most satisfying views of meteors are seen with the naked eye as they streak across the sky. Observing groups such as ALPO encourage astronomers to view meteor showers and submit their observations for review.

JUPITER

The largest planet and closest gas giant to Earth, Jupiter is easy to view. Its features are ever changing, making it a very interesting planet to watch. Jupiter is visible to the naked eye and resolves into a very small disk in good binoculars. Through a telescope, many of Jupiter's features are immediately apparent; unlike Venus, Jupiter gives few hazy or false readings. Jupiter's Great Red Spot, a huge swirling tornado or dust storm that has been raging on the planet for hundreds or even thousands of years, can be seen easily at powers of ×200 or so in nearly any telescope.

Jupiter appears to be divided into belts of varying hues. These belts darken and lighten unpredictably, and thus Jupiter never looks the same. The planet is nearly thirteen hundred times the size of Earth, and it is clear that some of its features are huge. The Great Red Spot, for example, is several times the size of Earth. Therefore, it is easy to see disturbances on the face of Jupiter. Dark spots erupt, swirling currents cross the face and disappear, and other fascinating planetary phenomena occur quite often, making Jupiter a challenging and exciting object for extended study.

Jupiter boasts many moons. Four of them (Io, Europa, Ganymede, and Callisto) are visible in binoculars, and several more are visible in larger telescopes. On its swing by Jupiter, *Voyager 2* tentatively identified at least fifteen moons, and there are perhaps several more, probably dragged into orbit around the huge planet from the asteroid belt. In a good telescope at high power, it is possible to watch the larger moons move across the face of Jupiter and to follow the shadows these moons cast as they eclipse small portions of the planet.

Courtesy NASA

Courtesy NASA

This photo of Jupiter, above, was taken by Voyager 1 in 1979. The Great Red Spot can be seen along the limb at the far right.

In a photo taken by Voyager 2, Saturn's rings are visibly transparent, indicating that they are not made of solid bands of matter. The shadow of the rings is apparent near the equator of the planet.

SATURN

Saturn's rings are its most famous feature, and one of the most spectacular sights in our solar system. The rings are made up of small particles orbiting the planet. They are not visible to the naked eye; the casual nighttime observer of Saturn sees only a medium-bright starlike object. Even binoculars won't resolve the rings—a minimum magnification of about ×20 is necessary even on the best of nights for seeing. However, large telescopes are necessary to truly observe Saturn and its features, particularly its moons.

Saturn is much less varied than its cousin, Jupiter, though it does show some markings on the disk. One interesting feature of Saturn is its shadow cast over the rings, and, similarly, shadows cast on Saturn by the rings. By examining these shadows, Saturn's orientation in relation to Earth can be easily discerned. Occultations of stars behind Saturn are also very interesting events to observe, because passing stars do not disappear behind the rings. This is proof that the rings are made up of small particles rather than solid belts of material.

Saturn, like Jupiter, has many moons, several of which are observable in larger telescopes. At least seventeen moons, the largest of which is Titan, have been confirmed orbiting around Saturn. Of these, six or seven are visible in telescopes of twelve inches or less. Many of these moons perform as would be expected of satellites, but it is clear from the odd behavior exhibited by a few that they were probably pulled into orbit around Saturn later in the planet's history, perhaps from the asteroid belt.

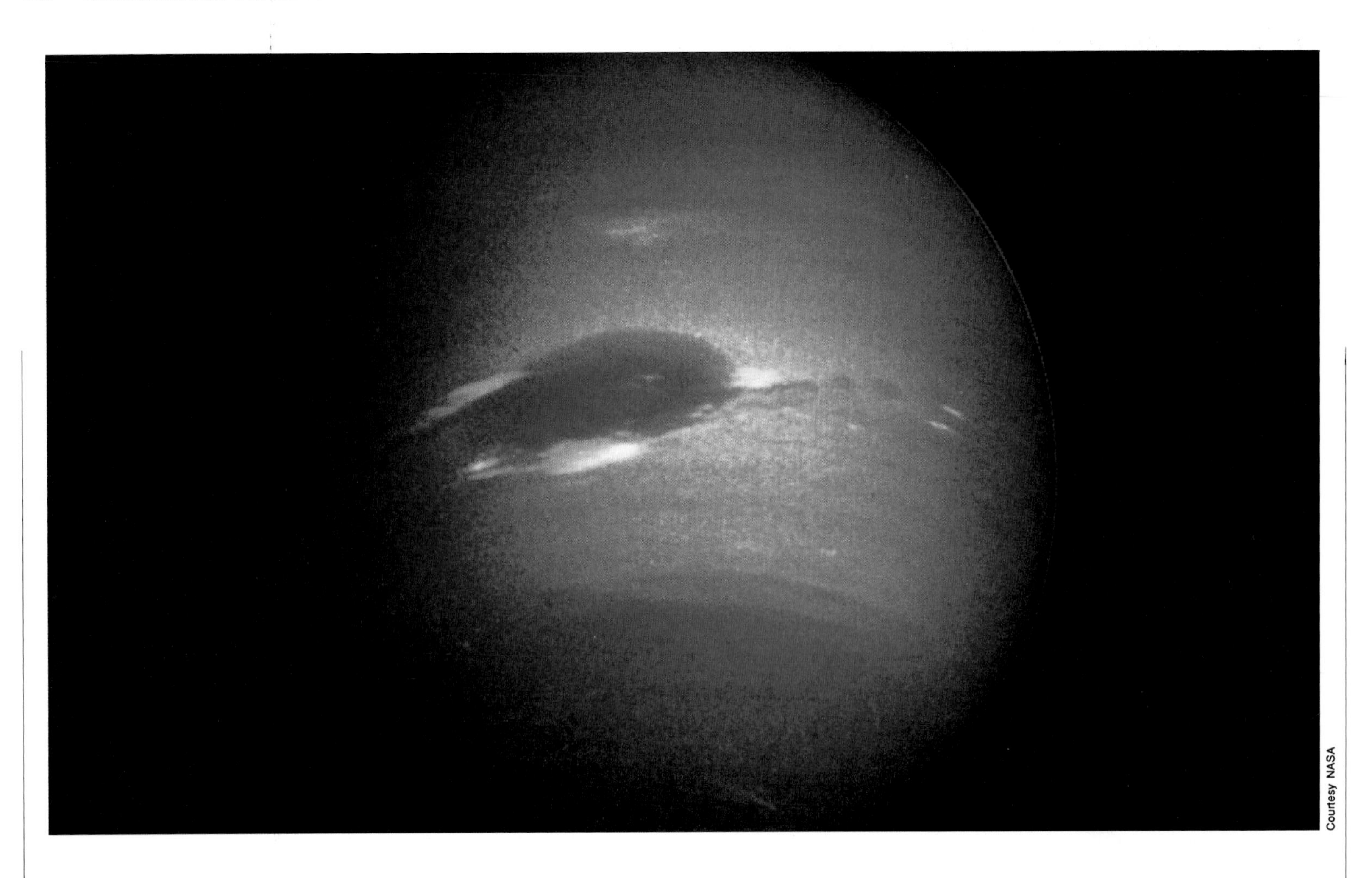

Courtesy NASA

THE OUTER PLANETS

Just as Jupiter and Saturn are very similar in size and makeup, so too are Uranus and Neptune. Both were telescopic discoveries, Uranus by William Herschel himself in 1781 and Neptune in 1846. Both are gas giants of similar size, about four times the diameter of Earth and one-third the diameter of neighboring Jupiter.

Uranus can just barely be seen with the naked eye, and it moves so slowly across the celestial sphere that until it was discovered by accident to be a planet, it was assumed to be one of the fixed stars. It can be resolved as a disk at $\times 40$ or greater, but it appears very dim because of its distance from Earth. Therefore, large apertures are necessary to pick up surface detail. Our knowledge of Uranus comes mainly from the *Voyager* probes. Though a couple of Uranus's moons are visible from Earth in large telescopes, several of the major features that were detected by *Voyager 2* cannot be seen even in the largest Earth telescopes. *Voyager 2* has confirmed the presence of at least a dozen moons, and it appears that there are faint rings, or portions of rings, around Uranus.

Neptune, roughly the same size as Uranus, moves even more slowly across the night sky and is considerably farther away from earth than Uranus. It can be seen in binoculars as a greenish star and resolved as a disk in a telescope of four inches of aperture or greater. However, little surface detail is apparent, and very few regular features have been catalogued on Neptune by observers

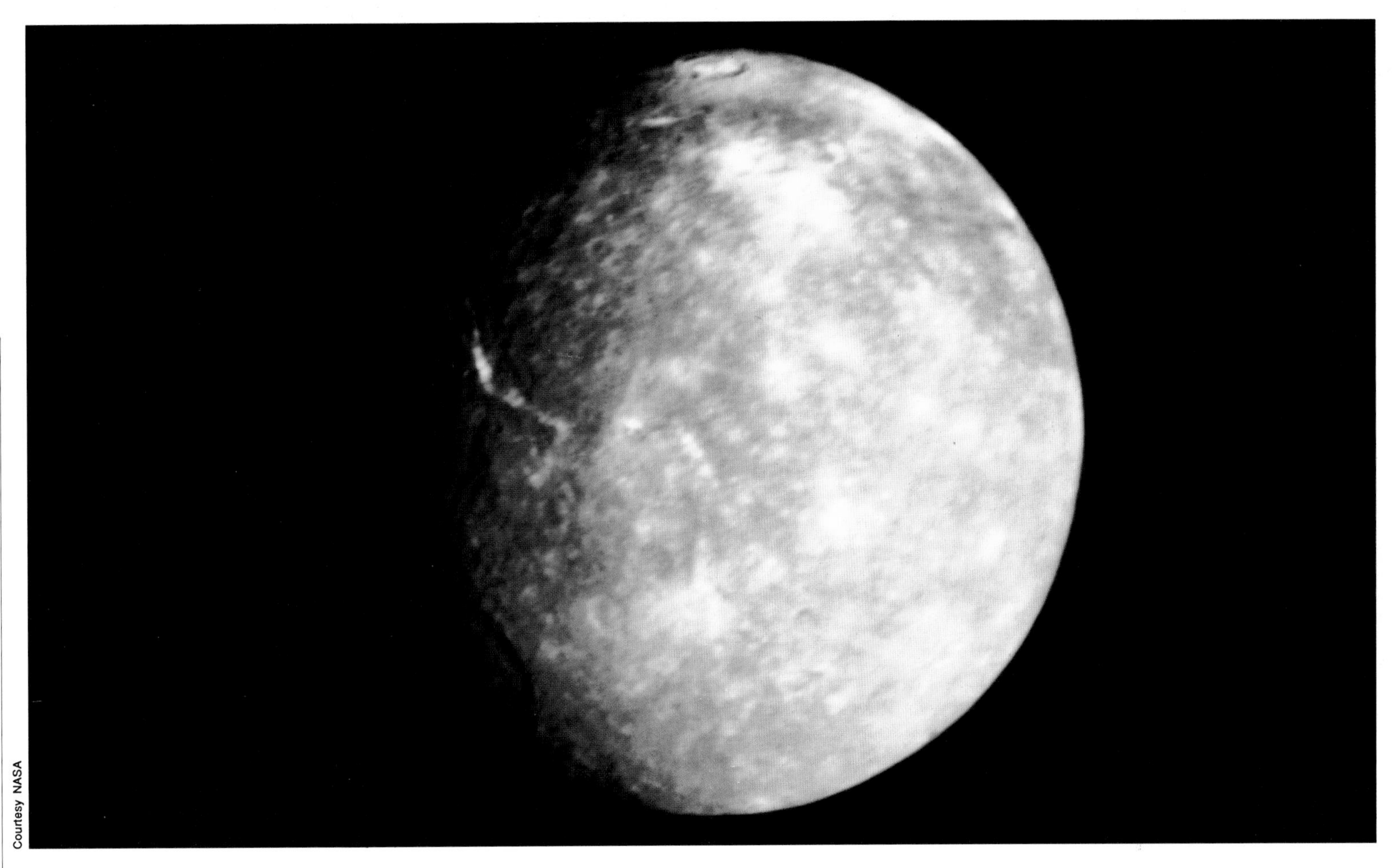

In the photo AT LEFT, a false-color image of Neptune was produced by combining images taken from ultraviolet, violet, and green filters on the wide angle camera of Voyager 2. This enhances the details of the cloud structure in Neptune's atmosphere. ABOVE, Voyager 2 captures Titania, one of the larger moons of Uranus, in a high resolution composite.

from Earth. *Voyager 2*'s recent passes of Neptune have granted Earth viewers a detailed first look at this cold planet. *Voyager 2* discovered that Neptune, far from being a frozen ball of ice, is in fact a very active planet with at least four defined rings, eight or more moons, and surface features not unlike those of Jupiter. Neptune has a large dark area, recently christened the "Great Dark Spot" (after Jupiter's Great Red Spot), that is visible in very large telescopes, as well as several other bright and dark spots large enough perhaps to be visible from Earth. When looking at Neptune armed with *Voyager 2*'s new information, however, amateur stargazers should be wary of "seeing" what they think is visible on Neptune rather than what actually is visible.

Pluto is an enigma. A small, frozen ball at the edge of the solar system, Pluto is not visible in any instrument with an aperture of less than eight inches, and it appears in large telescopes only as a faint star. Because of its very small size and extreme distance from Earth, Pluto is a most elusive target for many astronomers. One of the bizarre characteristics of Pluto is its irregular orbit—it is currently nearer to Earth than Neptune, and its orbit carries it into Neptune's orbit for a significant portion of its revolution about the sun. In 1978 a moon, Charon, was discovered, and it appears that Pluto and the large satellite Charon (which is almost half the size of Pluto) are more of a binary system than a traditional planet and satellite. Pluto is an object for the stargazer to add to a sighting list, but no meaningful observations of the planet can be made from Earth.

CHAPTER NINE

HUNTING FOR COMETS

Courtesy National Optical Astronomy Observatories

The most famous of periodic comets, Halley's, shown here in a long exposure photograph through star trails.

Comet hunting is a favorite activity for dedicated observers. Often, comets can be spotted with the naked eye or binoculars, and large telescopes are at a disadvantage in comet hunting because they cannot easily sweep large areas of the sky.

Comets are small masses of gas, dust, and ice that orbit the sun very irregularly. Some comets are visible with the naked eye, though most are objects for binoculars and telescopes. There are perhaps several hundred thousand comets in our solar system, though only a dozen or so are observable from Earth each year. Perhaps one every five years is visible to the naked eye, and only one in twenty years is bright and spectacular.

Comets consist of a nucleus of material, a cloud of melting ice and dust around the nucleus called the coma, and a tail, which is a trail of material from the coma. A comet becomes visible as it approaches the sun because it begins to vaporize from the sun's heat. The tail of a comet is almost always pointing away from the sun, because the solar wind (atomic particles "blowing" out from the sun) causes the tail to be pushed away. That is why a comet is always said to be facing the sun, though it does not actually travel in the direction it points.

Hunting for comets is a favorite activity of many stargazers, though it is not an easy task. Binoculars or small telescopes are best for comet hunting, and magnification is not as important as a rich field of view. The comet hunter must be very familiar with the night sky, as nebulas and star clusters may sometimes be mistaken for faint comets. However, a good knowledge of the skies and several star atlases handy for quick reference will help avoid such errors. Faint objects should be isolated in the field of view as much as possible, then the position should be found in relation to nearby marker stars. At this point, star charts and sky atlases should be consulted to rule out the possibility of a nebula, galaxy, or star cluster rather than a comet. Star charts will explain away almost any sighting of a potential comet, except for that one-in-a-million event when the object actually is a comet.

The best method of looking for comets is to slowly sweep over an area of sky, gazing with averted vision across the field of view. Observations should be carried out in very dark skies, as the contrast of light and dark will help make a faint comet more visible. It is better to survey small areas of the sky very carefully than to pass through huge chunks of the sky only cursorily.

There are several impediments to comet hunting. One is overcast skies—several consecutive days of sky cover can prevent sustained observations. Another problem is moonlight—the full moon often hinders viewing in a good portion of the sky because of the light it throws off. The most insidious problem in looking for comets is eyestrain, because of the extensive viewing that is required. To combat eyestrain, test the eyes at the end of each sweep of a band of sky and avoid hunting for more than an hour in a single evening.

© Alan McClure

© Alan McClure

These comet photos clearly show the anatomy of comets. The fiery head is called the coma, and the trail of gas and matter emanating from the coma is the tail. Note that the direction the comet "faces" is not the direction the comet is traveling, but rather the direction to the Sun (since the comet's tail opposes the Sun due to solar wind).

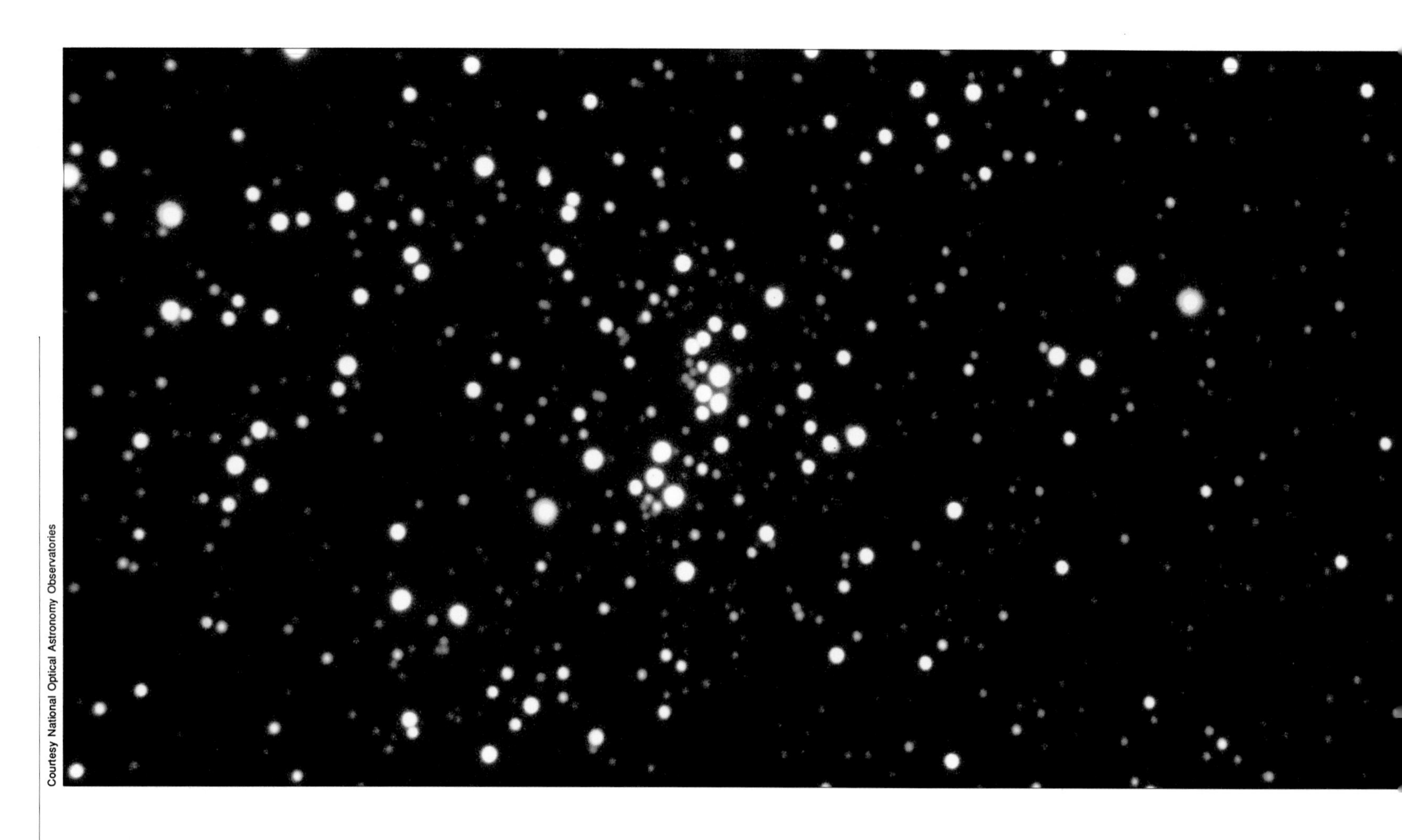

Courtesy National Optical Astronomy Observatories

VIEWING THE STARS

The sun is only one of several billion stars in the universe. Stars fill just about every bit of the night sky, though many are so faint as to be invisible in all but the largest telescopes. Stars like our sun group together in pairs called binaries; in groups called clusters; and in huge conglomerations of many millions called galaxies. All of these types of star groupings are visible in the night sky.

The stars are incredibly far apart. If we say that the sun is the size of a softball, Earth at this scale would be across a regular city street from the sun and only the size of a grain of sand. Pluto would be about six blocks away, and the nearest star to the sun, if the sun were in New York City, would be in St. Louis. At such distances, the stars can never be resolved as objects in even the largest telescopes.

Of the many stars in the sky, there are a few that are most interesting to the amateur observer. The brightest stars are the signposts of the sky, familiar landmarks on which to take one's bearings. Double stars, two (or more) stars so close together in the sky that they appear as one to the naked eye, are particularly interesting to separate under magnification. Variable stars, whose magnitudes vary for a number of reasons, are also objects of interest, both for observation and study. Finally, groups of stars such as clusters provide sometimes spectacular views and offer glimpses into other worlds. Other galactic phenomena such as novas and supernovas are exciting to observe, though they are seen very rarely.

The Aurora Borealis (or Northern Lights, pictured above) and the Aurora Australis (the Southern Lights) appear more spectacular as you observe closer to the Earth's magnetic poles. They are made up of charged particles that are excited when they strike the Earth's atmosphere, causing them to glow in the manner of a neon light.

1987B, and so on. The prefix SN stands for supernova, which is both an accurate and inaccurate name. Originally, supernovas were differentiated from novas by their magnitude: Supernovas are visible in daylight with the naked eye, while novas are only visible in the night sky. However, since novas and supernovas are the same types of events, they are now collectively referred to by the SN prefix.

Observing variable stars is one area in which amateur stargazers can make significant contributions to science. It requires describing any changes in a variable star's magnitude over a sustained observation period. This is accomplished by estimating the magnitude of a known variable by comparing its apparent magnitude at each observing session with the known magnitudes of nearby comparison stars.

Comparison stars are stars that are known to be unchanging in their magnitude and can thus be used profitably for visual comparisons. At first, most observers are only able to differentiate up to whole magnitude differences, but experienced variable star observers can accurately gauge varying brightness to one tenth of a level of magnitude. Practice and patience are the most important qualities for variable star observers, because the paramount consideration is the accuracy of the magnitude estimates. The American Association of Variable Star Observers (AAVSO) conducts this type of research and welcomes serious and experienced observers into its membership. The address can be found at the end of this book.

C H A P T E R   T E N

NEBULAS AND GALAXIES

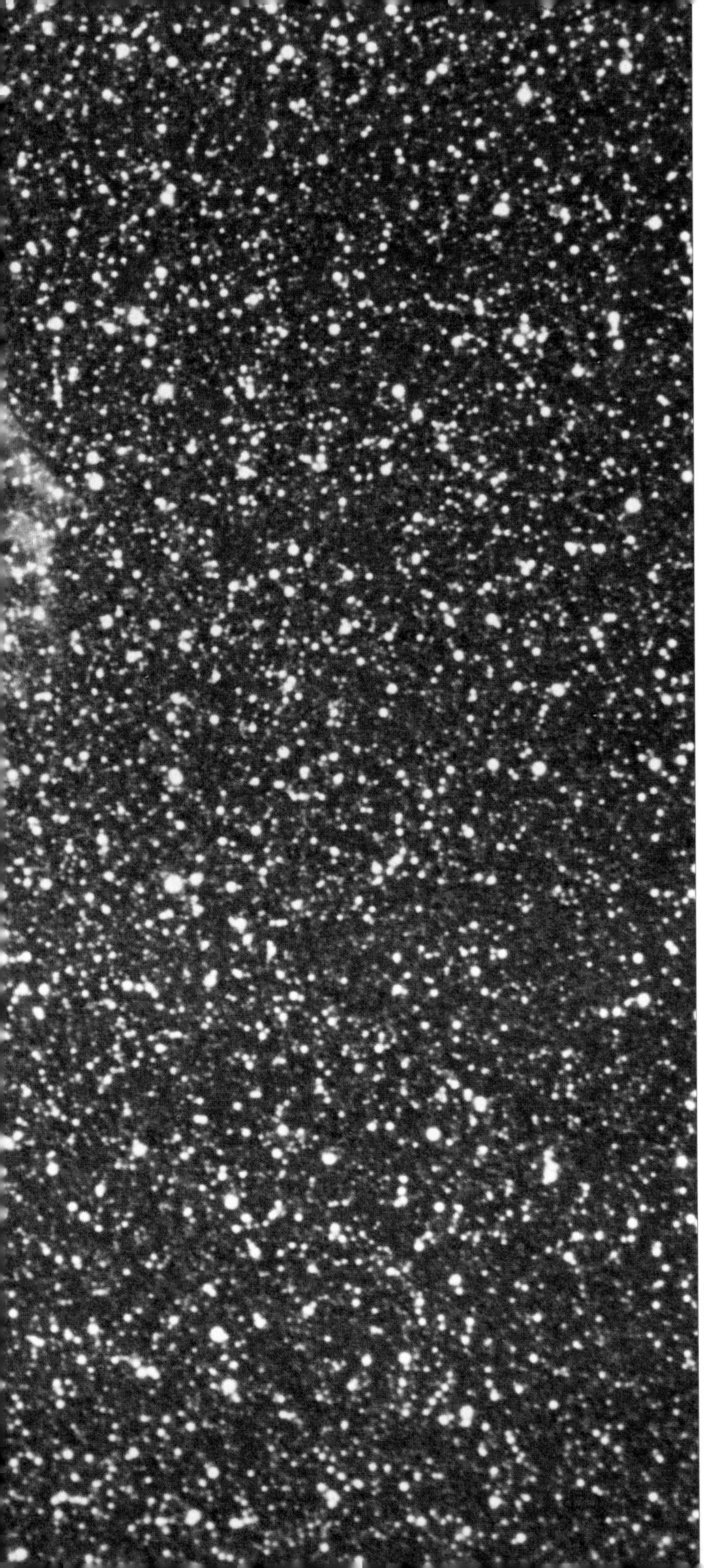

Courtesy Celestron International

This photo of the Veil Nebula was taken through a Celestron 5 1/2-inch (14-cm) Schmidt camera.

Courtesy Celestron International

The Whirlpool Galaxy and the Pinwheel Galaxy, ABOVE, *are spiral galaxies of the type Sc. "S" denotes a spiral galaxy, and "c" is a measure of the openness of the spiral arms, "a" being the most closed and "c" the most open. Our galaxy, the Milky Way, is another spiral galaxy, type Sb,* AT RIGHT. FOLLOWING PAGE, *the relative sizes and orbits of the outer planets in our solar system. The irregular orbit of Pluto is just one of its eccentricities.*

There is often some confusion regarding nebulas and galaxies. Both are huge groups of matter that can usually be seen only very vaguely even through telescopes (but there are a few galaxies visible with the naked eye). The difference between the two is this: A galaxy is the greatest conglomeration of stars that we can see, and galaxies encompass all other groupings. Portions of our own galaxy, the Milky Way, are visible with no optical aids—the spread of stars appears as a light, wide band crossing the sky. There are many, many galaxies in the universe, each containing many billions of stars.

Nebulas, on the other hand, are groupings of matter that exist inside a galaxy. There are several nebulas visible from medium to large telescopes, including the famous Crab Nebula and the oddly dark Horsehead Nebula. Nebulas consist of great clouds of particles or other types of matter that accumulate when a star explodes or is born. Often this matter takes on a distinct hue and shape and appears to glow faintly, perhaps with a faint star just visible near its center, behind the clouds of dust. In other cases, such as the Horsehead Nebula, the particles may be extremely dark and hide any light emitted rather than glowing themselves.

Galaxies are the neighborhoods of the universe, the areas where the stars seem to gather. Nebulas, on the other hand, signal the end of a star or the beginning of one or more. Both galaxies and nebulas are stunning to look at, though they actually are better objects for a camera, which can capture their light for extended periods, than for a telescope.

Courtesy Celestron International

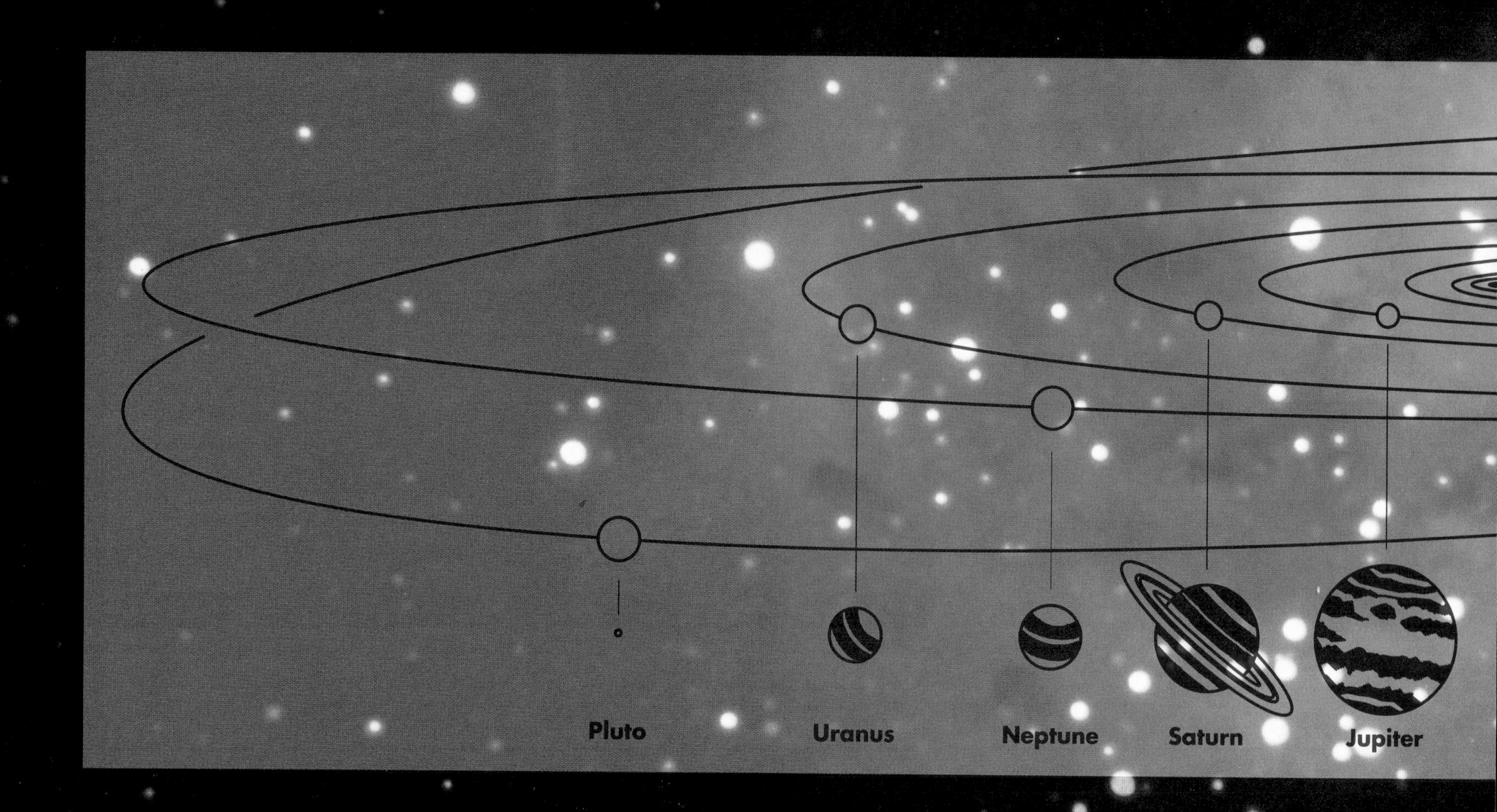
Pluto
Uranus
Neptune
Saturn
Jupiter

Venus
Mercury
Earth
Mars
© Anne Meskey

AT LEFT, a variety of man-made garbage (some of it still functional) is floating throughout our solar system. Like a soda can left at the summit of Mt. Everest, mankind's pollution leaves its mark on all frontiers. ABOVE, our knowledge of the universe is improving through the use of non-visual observations, such as radio telescopy, gamma-ray measurement, and collection of x-ray emission data from other celestial bodies.

MAN-MADE SATELLITES AND SPACE JUNK

If you tire of observing the natural wonders of the universe, why not try looking at the unnatural wonders, the newest members of the heavens. Man has, in the past thirty years, sent thousands of objects into space. Some are short-duration flyers, which usually don't stray too far from our atmosphere and don't remain in the sky for long. Others, like Skylab, go up and come down unexpectedly, or break up and become space junk, floating through space with no hope of being retrieved or disposed of. Still others travel so far away from Earth that they become hopelessly lost to observers. However, there remain many man-made objects in space that are easily visible, particularly for telescopic observers.

Jim Hale has shepherded the idea of observing man-made objects in space. His group, the Amateur Satellite Observers (ASO), meticulously combs the skies for them. Each month, Hale puts out a newsletter informing members of upcoming launches, satellites that will be visible, news on future satellites, and computer programs that will allow a computer-driven telescope to locate and track particular satellites. Hale scores real coups when he reports on military satellite launches, as most of the information about these is classified.

Man-made satellite observation has its drawbacks, however. For one thing, believing isn't always seeing. For instance, many observers reported seeing the Apollo missions land on the moon through their telescopes, a feat which would have been impossible. Another problem is that satellites sometimes are visible only at inconvenient or even impossible times and places. Finally, observing satellites with military uses, while challenging, is particularly frustrating because of the lack of data available for tracking.

However, the above caveats aside, observing man-made satellites can be very interesting and rewarding. And in the near future, with items such as the Hubble Space Telescope and the Space Station being readied for launch, the skies will only become more cluttered, and, for observers, more full of fascinating objects for viewing.

OBSERVING GROUPS

AMATEUR SATELLITE OBSERVERS (ASO)
HCR 65, Box 261-B
Kingston, AR 72742
(Offers a newsletter, *The Data Letter*)

AMERICAN ASSOCIATION OF VARIABLE STAR OBSERVERS (AAVSO)
25 Birch Street
Cambridge, MA 02138

AMERICAN METEOR SOCIETY
Department of Physics and Astronomy
State University of New York
Genesco, NY 14454

ASSOCIATION OF LUNAR AND PLANETARY OBSERVERS (ALPO)
P.O. Box 16131
San Francisco, CA 94116

SAN FRANCISCO SIDEWALK ASTRONOMERS
1801 Golden Gate Avenue
San Francisco, CA 94115

STRIKING SPARKS
Sonoma County Astronomical Society
P.O. Box 183
Santa Rosa, CA 95402

PUBLICATIONS

MERCURY MAGAZINE
Astronomical Society of the Pacific
1290 24th Avenue
San Francisco, CA 94122
(Magazine of observing group)

THE OBSERVER'S GUIDE
P.O. Box 35
Natrona Heights, PA 15065
(Magazine)

SKY PUBLISHING CORPORATION
49 Bay State Road
Cambridge, MA 02238
(Books, *Sky & Telescope*)

THE STARRY MESSENGER
P.O. Box 4823-N
Ithaca, NY 14852
(Magazine)

WILLMAN-BELL, INC.
P.O. Box 35025
Richmond, VA 23235
(Books)

TOURS AND TRAVEL

ECLIPSE ENTERPRISES, LTD.
1080 Fifth Avenue
New York, NY 10028
(Tours to observing sites)

VIRGINIA ROTH
Scientific Expeditions
211 East 43rd Street Suite 1404
New York, NY 10017
(Solar Eclipse Tours)

STAR HILL INN
Sapello, NM 87745
(Astronomical Retreat)

SOURCES

EQUIPMENT (INCLUDING TELESCOPES, OPTICS, CHARTS, MAPS, ACCESSORIES, SOFTWARE, AND MORE)

AD-LIBS ASTRONOMICS
2401 Tee Circle
Suites 105 and 106
Norman, OK 73069

ADORAMA
42 West 18th Street
New York, NY 10011

ARCHIVE PC
P.O. Box 59
Flanders, NJ 07836

ASH MANUFACTURING COMPANY, INC.
Box 312
Plainfield, IL 60544

ASTROMURALS
P.O. Box 7563
Washington, DC 20044

ASTRO-PHYSICS
7470 Forest Hills Road
Loves Park, IL 61111

ASTRO-TECH
101 West Main
P.O. Box 2001
Ardmore, OK 73402

ASTRO WORLD
5126 Belair Road
Baltimore, MD 21206

BOB'S ELECTRONIC SERVICE
7605 Deland Avenue
Fort Pierce, FL 34951

CALIFORNIA TELESCOPE COMPANY
P.O. Box 1338
Burbank, CA 91507

CELESTRON INTERNATIONAL
2835 Columbia Street
Torrance, CA 90503

CENTURY TELESCOPE
12555 Harbor Boulevard
Garden Grove, CA 92640

CHICAGO OPTICAL
P.O. Box 1361
Morton Grove, IL 60053

COSMIC CONNECTIONS
1460 North Farnsworth
Aurora, IL 60505

COULTER OPTICAL, INC.
P.O. Box K
Idyllwild, CA 92349-1107

CRITERION SCIENTIFIC INSTRUMENTS, INC.
620 Oakwood Avenue
West Hartford, CT 06110

D & G OPTICAL
6490 East Lemon Street
East Petersburg, PA 17520

EDMUND SCIENTIFIC CORPORATION
Edscorp Building
101 East Gloucester Pike
Barrington, NJ 08007

FARQUHAR GLOBES
5007 Warrington Avenue
Philadelphia, PA 19143

GALAXY OPTICS
P.O. Box 2045
Buena Vista, CO 81211

HANSEN PLANETARIUM PUBLICATIONS
1098 South 200 West
Salt Lake City, UT 84101

EDWIN HIRSCH
168 Lakeview Drive
RR 2
Tomkins Cove, NY 10986

JIM'S MOBILE IND.
1960 County Road 23
Evergreen, CO 80439

ROBERT T. LITTLE
P.O. Box E
Brooklyn, NY 11202

LORRAINE PRECISION OPTICS
1319 Libby Lane
New Richmond, OH 45157

© John Smith & Eddie Dymora/Texas Nautical Repair

HERBERT A. LUFT
P.O. Box 91
Oakland Gardens, NY 11364

LUMICON
2111 Research Drive, #5S
Livermore, CA 94550

MARTIN'S STAR TRACKER
3163 Walnut Street
Boulder, CO 80301

MEADE INSTRUMENTS CORPORATION
1675 Toronto Way
Costa Mesa, CA 92626

MMI CORPORATION
Dept. ST-89
2950 Wyman Parkway
Box 19907
Baltimore, MD 21211

NATIONAL CAMERA EXCHANGE
9300 Olson Memorial Highway
Golden Valley, MN 55427

NORTHERN LITES
801 Stanehill Place
RR 1
Victoria, BC V8X 3W9

NORTHERN SKY TELESCOPES
5667 Duluth Street
Golden Valley, MN 55422

KENNETH NOVAK & CO.
Box 68W
Ladysmith, WI 54848

OBSERVA-DOME LABORATORIES, INC.
371 Commerce Park Drive
Jackson, MS 39213

OPTIC-CRAFT MACHINING
33918 Macomb
Farmington, MI 48024

ORION TELESCOPE CENTER
421 Soquel Avenue
Dept. N
P.O. Box 1158
Santa Cruz, CA 95061

PARKS OPTICAL
270 Easy Street
Simi Valley, CA 93065

PAULI'S WHOLESALE OPTICS DIV.
29 Kingswood Road
Danbury, CT 06811

QUESTAR
P.O. Box C
New Hope, PA 18938

S & S OPTIKA, LTD.
5172 South Broadway
Englewood, CO 80110

SCIENTIA, INC.
1815 Landrake Road
Towson, MD 21204

SCOPE CITY
679 Easy Street
Simi Valley, CA 93065

SPECTRA
6631 Wilbur Avenue
Suite 30
Reseda, CA 91335

STAR LINER CO.
1106 South Columbus
Tucson, AZ 85711

STELLAR SOFTWARE
P.O. Box 10183
Berkeley, CA 94709

SUNDIALS INC.
Sawyer Passway
Fitchburg, MA 01420

TECTRON TELESCOPES
2111 Whitfield Park Avenue
Sarasota, FL 34243

TELESCOPICS
P.O. Box 98
La Canada, CA 91011

TELE VUE
20 Dexter Plaza
Pearl River, NY 10965

TEXAS NAUTICAL REPAIR
2129 Westheimer
Houston, TX 77098

THOUSAND OAKS OPTICAL
Box 248098
Farmington, MI 48322-8098

ROGER W. TUTHILL, INC.
11 Tanglewood Lane
Mountainside, NJ 07092

UNITRON INSTRUMENTS
175 Express Street
Plainview, NY 11803

UNIVERSITY OPTICS
P.O. Box 1205
Ann Arbor, MI 48106

VERNONSCOPE & CO.
Candor, NY 13743

WHOLESALE OPTICS OF PENNSYLVANIA
Box 15
Sterling, PA 18463-0015

ZEPHYR SERVICES
1900 Murray Avenue
Dept. A
Pittsburgh, PA 15217

© Gary Ladd

PLANETARIUMS

ADLER PLANETARIUM
1300 South Lake Shore Drive
Chicago, IL 60605

ALBERT EINSTEIN PLANETARIUM
National Air and Space Museum
Smithsonian Institution
Washington, DC 20560

ALLEGHENY OBSERVATORY
Department of Physics and Astronomy
University of Pittsburgh
Pittsburgh, PA 15214

AMERICAN MUSEUM, HAYDEN PLANETARIUM
81st Street at Central Park West
New York, NY 10024

BUHL PLANETARIUM
Allegheny Square
Pittsburgh, PA 15212

CHARLES HAYDEN PLANETARIUM
Museum of Science
Boston, MA 02114

DAVIS PLANETARIUM
Maryland Science Center
601 Light Street
Baltimore, MD 21230

FELS PLANETARIUM
20th and Parkway
Philadelphia, PA 19103

FERNBANK SCIENCE CENTER
Jim Cherry Memorial Planetarium
156 Heaton Park Drive
Atlanta, GA 30307

GATES PLANETARIUM
Colorado Boulevard and Montview
Denver, CO 80205

GEORGE T. HANSEN PLANETARIUM
1098 South 200 West
Salt Lake City, UT 84101

GRACE H. FLANDRAU PLANETARIUM
University of Arizona
Tucson, AZ 85721

GRIFFITH OBSERVATORY
2800 East Observatory Road
Los Angeles, CA 90027

HALE OBSERVATORIES
(Mt. Wilson and Mt. Palomar)
California Institute of Technology
Pasadena, CA 91101

HARVARD-SMITHSONIAN CENTER FOR ASTROPHYSICS
60 Garden Street
Cambridge, MA 02138

KITT PEAK NATIONAL OBSERVATORY
P.O. Box 26732
Tucson, AZ 85726

LICK OBSERVATORY
Department of Astronomy
University of California at Santa Cruz
Santa Cruz, CA 95064

LOUISIANA ARTS AND SCIENCES CENTER PLANETARIUM
503 North Boulevard
Baton Rouge, LA 70802

McDONALD OBSERVATORY
Department of Astronomy
University of Texas
Austin, TX 78712

McDONNELL PLANETARIUM
5100 Clayton Avenue
St. Louis, MO 63110

MOREHEAD PLANETARIUM
University of North Carolina at Chapel Hill
Chapel Hill, NC 27599-3480

MORRISON PLANETARIUM
Academy of Sciences
San Francisco, CA 94118

NATIONAL RADIO AND IONOSPHERIC OBSERVATORY (Puerto Rico)
P.O. Box 995
Arecibo, PR 00612

NATIONAL RADIO ASTRONOMY OBSERVATORY
Charlottesville, VA 22901

R. C. DAVIS PLANETARIUM
P.O. Box 288
Jackson, MS 39205

REUBEN H. FLEET SPACE THEATER
1875 El Prado
San Diego, CA 92103

SPACE TRANSIT PLANETARIUM
Rochester Museum and Science Center
663 East Avenue
Rochester, NY 14607

SPROUL OBSERVATORY
Swarthmore College
Swarthmore, PA 19081

STRASENBURGH PLANETARIUM
Rochester Museum and Science Center
663 East Avenue
Rochester, NY 14607

TALBERT AND LEOTA ABRAMS PLANETARIUM
Science Road
Michigan State University
East Lansing, MI 48824

UNITED STATES NAVAL OBSERVATORY
Washington, DC 20390

VANDERBILT PLANETARIUM
178 Little Neck Road
Centerport, NY 11721

YERKES OBSERVATORY
University of Chicago
Chicago, IL 60637

INDEX

Additional Photography Credits

Courtesy Celestron International: 20-21, 132-133
© Geoff Chester: 64-65
© George East: 70-71
Courtesy National Optical Astronomy Observatories: 2-3